Wildlife of Rivers and Canals

Wildlife of Rivers and Canals

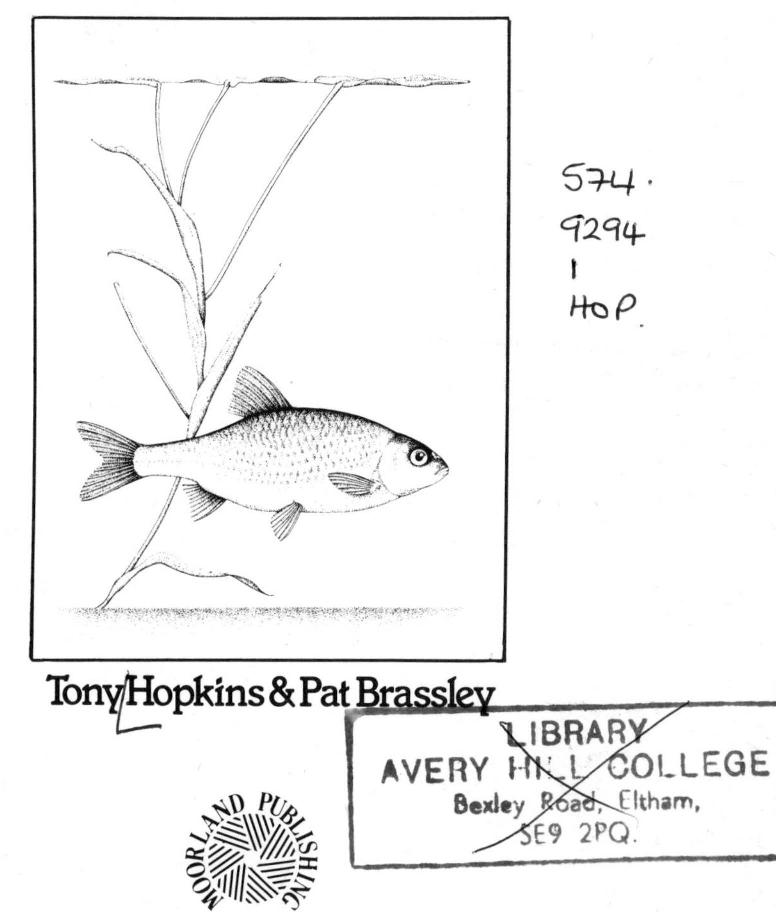

Tony Hopkins & Pat Brassley

MOORLAND PUBLISHING

British Library Cataloguing in Publication Data

Hopkins, Anthony J.
Wildlife of rivers and canals
1. Fresh-water fauna — Great Britain
2. Fresh-water flora — Great Britain
I. Title II. Brassley, Patricia
574.92'9'41 QH137

ISBN 0 86190 061 8

Printed by Butler and Tanner Ltd, Frome
for Moorland Publishing Company Ltd,
PO Box 2, 9-11 Station Street, Ashbourne,
Derbyshire, DE6 1DZ, England

CONTENTS

LIST OF ILLUSTRATIONS

Acknowledgements for Colour Plates

N. Dunstone: mink
B. Gadsby: kingfisher, moorhen
S. Hilton: cyclops, fishlouse
A.J. Hopkins: dog rose, elephant hawk moth larva, monkey flower, orange tip butterfly, scarlet tiger moth
P. Kirk: dragonfly, grass snake, great diving beetle, reed warbler
G. Nobes (Press-Tige Pictures): heron (dust jacket)

Preface

Most books about freshwater only deal with what is happening below the surface, and even then the highly-adapted creatures of fast-flowing rivers are often given more than their fair share of attention because they provide useful models for the study of ecological systems. Yet most people live in lowland Britain near slow-flowing waters, and the majority visit canals or rivers without any specialised equipment — not even a pond net.

This book is intended as a simple guide to the wildlife of these interesting and rich habitats, for the use of amateur naturalists on riverside walks or families on a narrow-boat holiday, who want to know about the water and what lives alongside it. The emphasis is on what can be seen above the surface, but account has to be taken of what happens beneath since one is dependent upon the other.

Wherever possible, individual chapters are organised in such a way that they emphasise the most interesting or most important aspects of waterway ecology. This means that semi-aquatic birds and mammals are considered individually, while those associated with banks and hedgerows are taken a family at a time. Insects are given special emphasis because of their diversity and the fact that few other books describe them adequately. Many of the smaller invertebrates do not have English names because they have not attracted enough public attention, but all plants and animals have a double (binomial) Latin name, which functions in much the same way as a surname followed by a christian name. Their length and complexity is sometimes daunting but they are always worth remembering, since English names can vary from place to place and year to year.

Chapters 1, 5, 6, 7, 8, 10 and 11 were written by Tony Hopkins, while Pat Brassley was responsible for chapters 2, 3, 4, 9 and 12.

All the illustrations, diagrams and black and white photographs are by Tony Hopkins.

The Oxford Canal at Napton in Warwickshire

1 *Introduction*

Rivers and canals lattice the countryside in such a way that it is difficult to avoid them. They share communities of plants and animals that make them unique, and yet subtle variations make a river different from a canal, and often one section of a river quite different from another.

The close of the ice-age about 12,000 years ago left a network of rivers to mould the lowland landscape of Britain. For many thousands of years after this they also controlled the destiny of settlers who had moved north across mainland Europe: first the Mesolithic hunter-gatherers, then Neolithic farmers. Rivers began as natural obstacles, then became important routes into the interior, and as the human population expanded they became navigation ways and the focus for settlement and commerce.

Navigable rivers provided an arterial system through lowland Britain, but as trade flourished in the eighteenth and nineteenth centuries artificial waterways were built to link the main systems. These canals became the capillaries for the industrial revolution, feeding factories with raw materials and filtering manufactured goods through to estuaries and ports. Rivers and canals shared this important role in history for a relatively short time: the ties that united them were of function, and once the railways had captured most of the transport market canals began a century of decline.

River wildlife had arrived shortly after the ice-age, as the glaciers diminished and the melt-water established a wide drainage system across Northern Europe. Animals and plants could spread uninterrupted along these new waterways, which were both fertile and free from competition. The rapid colonisation of Britain ended when the sea finally channelled out the Straits of Dover, but by then most of the important elements of river life were already here.

The climate has deteriorated since then, passing from the warm and wet Atlantic period through the warm sunny Sub-boreal to the cool wet climate we know today. The impact of man on the landscape has increased too, so it would be surprising if British water systems were as prolific today as they used to be. Certainly many species —

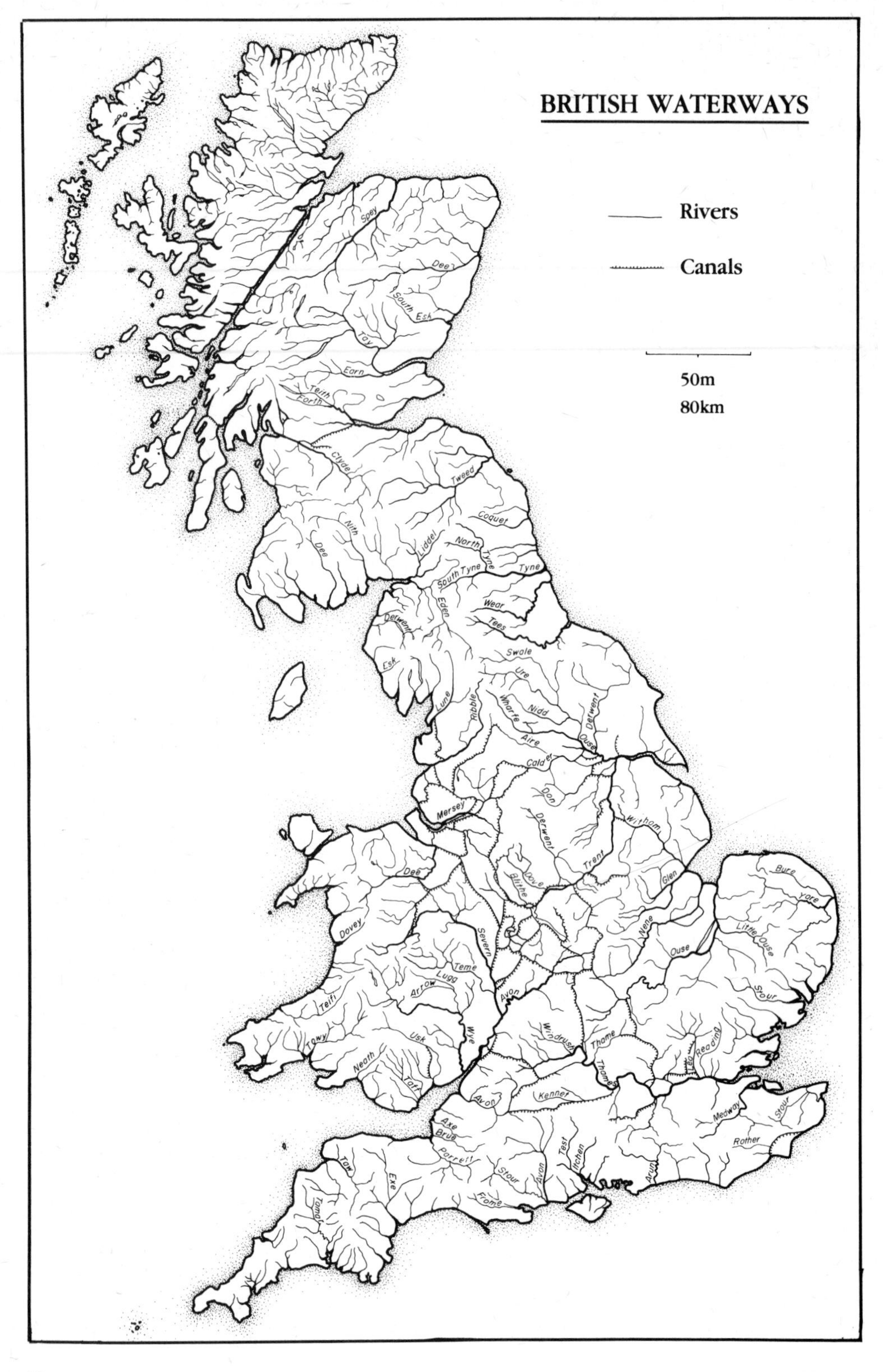
BRITISH WATERWAYS
Rivers
Canals
50m
80km
Spey
Dee
South Esk
Tay
Earn
Teith
Forth
Clyde
Tweed
Coquet
Nith
Dee
Liddel
North Tyne
South Tyne
Tyne
Eden
Wear
Derwent
Tees
Swale
Esk
Ure
Lune
Ribble
Wharfe
Nidd
Derwent
Aire
Ouse
Calder
Don
Mersey
Derwent
Witham
Dee
Trent
Dove
Blithe
Glen
Bure
Yare
Dovey
Severn
Nene
Little Ouse
Ouse
Teme
Lugg
Arrow
Avon
Stour
Teifi
Windrush
Thame
Wye
Towy
Usk
Lea
Neath
Taff
Thames
Avon
Kennet
Medway
Stour
Axe
Brue
Rother
Parrett
Test
Itchen
Taw
Exe
Stour
Avon
Arun
Frome
Tamar

for example the beaver — have become extinct in quite recent times, but to counter-balance this there have been new arrivals, either expanding naturally or through accidental introductions by man.

A recent colonist of slow rivers, the great crested grebe, Podiceps cristata, *demonstrates the continuing process of retraction and expansion in wildlife populations*

When the first true canals were built towards the end of the eighteenth century they provided a potential wildlife habitat, but it is impossible to assess how quickly they were colonised or how verge and water management worked against the establishment of plant and animal communities. Although traffic was regular and heavy there was no pollution. Boats travelled slowly and there was no damage to banks, and no propeller blades to churn the bottom mud. At the same time, a great deal of care was taken to keep the vegetation in check, and the decline of water traffic and the dereliction of many canals must have brought a wide and rapid expansion of waterside life.

The long period of dereliction highlights an important factor in the difference between natural drainage systems and artificial cuts. Rivers could maintain themselves without any interference and from the wildlife point of view the less management they receive today the better the habitat that remains. But canals need to be maintained and controlled and the waterlife that flourished in the derelict years brought about its own downfall by choking the banks and allowing shrubs to establish themselves. Basins dried out, cracked and were filled in, and by the time the demand for leisure had opened the inland waterways to new prosperity it was already too late to save many of them. Canals that had not destroyed themselves were filled in by farmers eager to reclaim land.

The plains of lowland Britain are drained by the same rivers that carried the ice melt-water. Unfortunately, rivers are unreliable and often flood their banks during the winter rains. The importance of land drainage brought a reprieve for some canals which have been incorporated into regional drainage schemes. Many new canals or dykes have been built for drainage rather than navigation, and many rivers have been 'canalised' to improve flow and reliability. The whole effect has been to draw rivers and canals together in an interrelated system, with some loss to their individuality. The contrast between natural rivers and traditional canals is still very marked, however, and is emphasised by the different influences involved. The ability of rivers to change overnight into torrents of flood-water is contrasted by the constancy of canal levels, which have a regulated input. River water flows swiftly, even in navigable lowland areas, while canal water makes slow progress, held up by lock gates and the need to keep levels constant. Verges of canals are remarkably consistent too; the banks, towpaths and hedgerows being recognisable anywhere in the country, while rivers may pass through meadows or woodland and display a wide variety of

Ash trees, forming a canopy at the entrance to the Newbould tunnel, on the Grand Union Canal near Rugby

wildlife over a short reach.

In spite of these contrasts, rivers, canals and their hybrids have far more in common than with other similar habitats. None is very much more than a ribbon of water, and a system of a verge, bank and water edge, stretched for mile upon mile across the countryside and providing an uninterrupted highway for whatever animals and plants are on the move. Man may have adapted the waterways to his own purpose, but they remain the birthright of water voles and whirligigs rather than anglers and holidaymakers.

The Background

2 *Past and Present*

Our river and canal systems have developed very differently, rivers having natural origins and canals artificially cut. From this stems much of the difference in the habitats and wildlife. However the development of many of the canals of central England has been inseparably linked with rivers as indicated by the names of some of them — the Trent and Mersey, the Thames and Severn and the Kennet and Avon, all connect ports across the country. Other canals use rivers, such as the Soar in Leicestershire, as parts of the canal system and often as an important water source.

The Development of Rivers

At its simplest the development of a river depends on two main factors, a gradient and an adequate source of water from rain, an overflowing lake or springs. The hills cause rain-bearing clouds to shed their load by the rise in altitude; thus the west side of the Pennines, the first rising ground encountered by prevailing south-westerly winds, is wetter than the eastern side. The rain falls onto the ground, moves downhill and where hard rock forms the surface it may remain as a sheet, picking up loose rocks and other debris as it goes. Once soft rock or soil is encountered, the flow may coalesce into rivulets. The debris carried by the water will help to gouge out soft rock and gradually a course, used by most of the water for much of the time, will develop and become established. The amount of water flowing through a channel will depend on the gradient of the slope and the size of the catchment area. Gradually the water channel will create its own valley on the slope by eroding material from the sides as well as by down cutting. If there is no lateral erosion a gorge may be formed, depending on the type of rock.

Looking at the patterns of present day rivers there are anomalies which can only be explained by looking at other factors, which have operated in geological time. Mountain-building phases may have produced one pattern for a river and its tributaries, only for a later upheaval to disrupt it. Variations in climate have produced arid periods, torrential rains and ice ages each lasting thousands of years.

The evidence that conditions and the rivers they created have varied throughout geological time is often easily seen. The river terraces of the Thames indicate at least a dozen well marked changes of sea level in the London Basin since the early Pleistocene, 2.5 million years ago. There are also buried channels and evidence that ice diverted the Thames from the Vale of St Albans, southwards to its present course. The 183m (600ft) sea level of the early Pleistocene can be traced from the chalk cliffs in south-east England to Lynton in North Devon. The subsequent falls in sea level are marked by a series of coastal wave-cut platforms corresponding to the river terraces. The sand and gravel deposits of the Thames and the Trent and the width of their valleys indicate that the rivers were once much larger. Sand and gravel deposits can also indicate ancient rivers long since lost, such as the one which laid down the pebble bed deposits in East Devon.

Glaciation, in particular, altered the pattern significantly in several ways. It was also the most recent major factor, although the south of England escaped many of the effects. Glaciers deposited debris, which dammed river valleys so that when warmer weather returned, lakes were created, as in the Lake District. The lakes may have spilled over in new places to form streams. Valleys became U-shaped as a result of the grinding action of the glaciers and many, which were widened and deepened by meltwater, now contain only small streams.

Alder trees and willows line the fields close to the River Teme, near Leintwardine on the Hereford/Shropshire border

Rock type can also have a significant effect on the course of a river. Sandstones, mudstones and shales erode easily, providing little resistance to the flow and enabling a channel to be cut through them. Limestone and chalk can be dissolved, causing streams to disappear down 'swallow holes' emerging yards or miles away where the limestone rests on harder rock. Granite and gritstone are hard rocks and erosion is slow, although the presence of weaker elements or the way the rock was formed may accelerate it. Some of the more interesting and attractive features of rivers depend on the juxtaposition of two types of rock, one hard, one soft, resulting in waterfalls, deep pools and acute bends.

The type of rock also affects the amount and effect of material carried by a river and the chemical characteristics of the water. For example large boulders or blocks might be produced by the erosion of conglomerate rock; they will move by sliding or rolling along the bed of the river, particularly when it is in spate. These boulders will crush other rocks, grind the bedrock and become rounded as their own corners are knocked off, gradually reducing the size of pebbles carried by the river. Large boulders are found mainly in the fast flowing upper reaches of streams. Pebbles, derived from the bedrock, bounce along spending most of their time in the water, but touching the bed occasionally. They can also reduce the size of other pebbles and themselves, depending on the water speed and force of the bouncing. Finer particles, derived from sandstones, for example, are carried in suspension and are deposited only when the water speed slows at a waterfall or bend. If there are a lot of suspended solids the visibility is affected and predatory fish cannot see their prey. Another way a river can carry material is in chemical solution and water flowing through chalk or limestone dissolves lime and the alkalinity (pH) is affected. Acid rocks will have the opposite effect and the resultant weak acid solution may further aid erosion by dissolving the cementing substances in sandstones, conglomerates or other rocks. Plants and animals are affected by the acidity or alkalinity of the water, although some rivers, like the Derbyshire Derwent, receive water from both acid and basic rocks and the effects may be neutralised.

Many people may be familiar with a particular stretch of a river, without realising that each river has several phases, although it may pass quickly through them. The main determining factor is the overall gradient, that is the height of the hills where the river starts, related to the distance from there to the sea. Rivers in parts of Scotland, northern and south-western England flow fairly directly from the hills to the sea. By contrast, in the Midlands and south of England where the hills are lower, the rivers wander on their way to the sea. The fall in metres per kilometre (feet per mile) is very variable; for the River Tees it is approximately six metres per kilometre (thirty feet per mile) whereas for the Hampshire Avon it is nearer one metre per kilometre (five feet per mile).

The various phases have been classified by different people in different ways: sometimes by physical features, for instance the speed or turbulence of the water, substrate, or the amount of silt or by botanical, zoological or geographical characteristics in any combination. An example of a physical/geographical classification is:

Mountain phase	Steep gradient. Erosion of material high, because of velocity of water and high carrying capacity.
Foothills phase	Break in gradient results in deposition of large material as velocity decreases.
Plains phase	Flatter land, water speed further reduced, silt, sand and gravel deposited. Lateral movement widens valley, creates oxbows in course. Bare rock is rare.
Estuary	Marine/brackish conditions.

The River Wye at Symonds Yat in the Forest of Dean

Rivers and Their Uses

To the early settlers rivers were the way of reaching into the heart of the country. They could also be boundaries, a danger when flooding, barriers and sources of disease. The benefits, including water for stock and people, fish (an important protein source in winter) and the means of disposing of rubbish and sewage, must have outweighed the problems. Gradually settlements grew up by rivers, often at important bridging points or where they could be utilised for transport or power. The potential for power provided by some rivers was harnessed for corn mills. Later in the eighteenth and nineteenth centuries the huge mills of the Pennine valleys in Yorkshire and Lancashire capitalised on the power and used the water for fulling (cleansing and thickening) cloth. River water was also controlled to produce grass for stock early in the year in managed water meadows.

Fords, bridges and the deepening of pools were probably the first artificial interruptions to the flow, followed by weirs, flood banks and mill leats providing greater control of the water, diverting it in directions more useful for traffic or less dangerous to nearby settlements. These man-made changes alter the gradient and flow and in turn change the wildlife habitats.

It is thought that the Romans improved on the existing river pattern by making several cuts such as the Foss Dyke from the River Trent to the River Witham at Lincoln, Car Dyke from Lincoln to Peterborough and Itchen Dyke which made Winchester accessible to the south coast. There is little evidence to show that much happened for a thousand years and it is likely that the dykes fell into disrepair. However, gradually, with the increasing need for fuel, food and goods, the poor state of the roads and the expense of land travel, the fifteenth and sixteenth centuries saw a great increase in works to improve river navigation. Thus from mediaeval times the major rivers of England carried a considerable trade, reaching towns far inland.

The major systems were the Thames, Great Ouse, Severn and Trent connecting with coastal traffic around their

estuaries. Only the Thames and Ouse are well documented but other busy river navigation systems did exist, including one on the Somerset levels serving Langport, Ilchester and Glastonbury Abbey. The Great Ouse, though shorter than both the Trent and the Severn, served eight counties and was navigable for ships of fifteen tonnes as far as Bedford and its tributaries served other important places, including Cambridge. Although it is known that the Severn was navigable to Welshpool in the sixteenth century, little is known of its trade, but by the seventeenth century it was a prosperous thoroughfare for coal and local manufactured goods. However, there is evidence that in dry spells even the Severn could be unnavigable for long periods and this was exacerbated by the actions of various landowners taking water and erecting weirs along its course. Several acts were passed in the reigns of Henry VIII and Elizabeth I relating to these obstructions and also to the heavy tolls imposed by the larger towns and cities. Conflicts with riparian owners were obviously a major difficulty and the Earls and Countesses of Devon erected weirs which obstructed the Exe below Exeter leading eventually to the construction of the Exeter Canal, opened in 1567. The first ship canal in Britain, it introduced pound locks, which already existed on the Continent and its construction was an early example of the local co-operation needed to initiate such a project.

During the seventeenth century more rivers were made navigable including the Soar in 1634 and some such as the Hampshire Avon, Wey Navigation and Great Ouse, were canalised. In some cases the work was simply widening and deepening the waterway, while in others it involved the construction of flash or pound locks or reservoirs to maintain the water level. Flash locks held water back until a boat was stranded, when the weir was opened by the removal of a section. Delays were caused because of the bargaining over the price of the water or, on a busy stretch, the need for the water to build up behind the weir. Barges and sailing ships were used and both were towed by men rather than animals, because there was usually no towpath and there were many obstacles to surmount, for instance trees, sidestreams and mills. The wind was unreliable and sails were rarely used except on wide rivers so men were still towing boats on rivers in the early nineteenth century.

Despite these drawbacks the use of waterways increased; before 1660 1,102km (685 miles) were navigable and this rose to 1,866km (1160 miles) by 1724, so that much of England was within 24km (15 miles) of a navigable waterway. This improvement influenced patterns of traffic, so that for example York's coal came by sea and river from Newcastle, rather than from the nearby West Riding from where only land transport was available. In the early eighteenth century the West Riding cloth

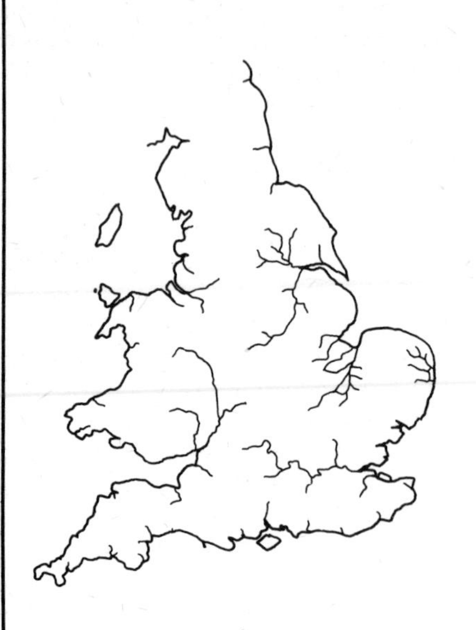

1760

1860

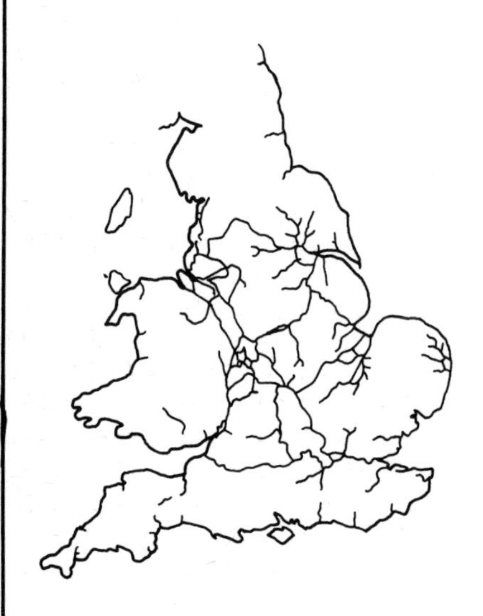

1980

Evolution and change in the navigable waterways of England and Wales, over three centuries between 1760 and 1980

Scale:

100m
(160km)

industry and the cutlery industry in Sheffield benefitted from the Aire and Calder Navigation and the Don Navigation respectively and corn for London came from the Great Ouse hinterland via Lynn (now King's Lynn).

In the mid-eighteenth century the situation was much the same, river transport was much cheaper and more efficient than the use of roads. One horse could draw 2 tonnes on the road, but 50-100 tonnes on water, and the roads were rough, sometimes impassable in the winter and robberies were frequent. However, river transport still had its difficulties including water shortages, haulage, rights of passage, delays while bargaining for water and the necessity to get goods and fuel to and from the nearest rivers. One of the problems of the transference of cargoes onto seagoing vessels near river mouths was the dumping of ballast which formed banks and prevented navigation. The clearance of silt deposited naturally, especially around man-made obstacles including mill weirs and fish traps, was very expensive. These problems led to a decline in inland water-borne trade and as a result towns were less inclined and less able to spend money clearing the silt and obstacles and so the fall in trade grew.

The Rise of Canals

Gradually the idea of totally new cuts or canals became popular and as canals were built so the length of navigable river declined. The first long distance canal was the Newry, 30km (18½ miles) long, on the east coast of Ulster, opened in 1742, followed by the St Helens Canal, on the north bank of the Mersey, which was completed in 1761. It was built by Thomas Berry as a by-pass for the Sankey Brook. In 1759 work began on the first well-known canal, the Bridgewater Canal, planned by James Brindley, a mill-wright, son of a Derbyshire farmer. The Bridgewater Canal, named after the third Duke, carried coal, loaded underground, directly from his mines at Worsley, into Manchester, where other fuels were scarce or costly when carried by road. The first section, completed in 1761, included the Barton aqueduct which was 183m (600ft) long and had three arches spanning the River Irwell. The canal was 5.5m (18ft) wide and 1.4m (4ft 6in) deep and in addition to the huge weight of water the aqueduct also had to bear that of the puddled clay which sealed the stonework. It was hailed as a wonder and lasted in good condition for 132 years until it was replaced when the Manchester Ship Canal was built.

The success of the Bridgewater Canal prompted others into forming companies to finance canal building, most of which was privately sponsored. An exception was the government help of £50,000 for the Forth and Clyde Canal (1784) which was recovered and reused for the Crinan Canal in 1799. Canal development was essentially conceived on a local basis; for example, pottery owners under the leadership of Josiah Wedgwood promoted the Trent and Mersey Canal and copper mine owners built the Tavistock Canal in Devon. Indeed for any group of people who needed raw materials, including fuel, and wanted to distribute their products, including manufactured goods and food, canals were an attractive proposition.

The companies had to obtain Acts of Parliament to cover the purchase of land, road crossings etc, although for small branches or canals where the promoters owned all the land, for example the Torrington Canal in Devon, this was not necessary. The essentially local, disjointed and selfish approach to canal construction led to many problems, some of which reduced the competitiveness of canals when the railways came. They included a lack of uniformity in channel size and depth, lock size and variation in tolls. There were broad canals, including the Kennet and Avon, Leeds and Liverpool and the Bridgewater, where boats between 17 and 24m (55 and 80ft) long and 3.6 and 6.4m (12 and 21ft) wide could be accommodated. Others such as the Chesterfield and Erewash are broad canals near the river but the locks are narrower and often shorter away from the junction. Narrow canals are only open to craft up to 21m (70ft) long and 2m (6ft 10in) wide. Another problem was the disputes over water supply when canals were joined. The stop lock at Autherley Junction, where the Shropshire Union Canal (formerly the Birmingham and Liverpool Junction) joins the Staffordshire and Worcestershire Canal, prevents water flowing out of the older waterway. The Worcester and Birmingham Canal and the Birmingham Canal Navigations (BCN) were separated at Gas Street Basin by the Worcester Bar and goods had to be trans-shipped across it, until eventually the two companies agreed to a stop lock. However, the individuality of the canals, the use of local materials and the types of locks, bridges and canalside buildings are all part of their attraction to the holiday maker and student alike. These differences also provide the diversity of wildlife habitats. The length of time taken to finish canals varied widely and depended on local money and enthusiasm; indeed the Leeds and Liverpool Canal took 40 years to complete. Engineers might change, dimensions could alter, innovations were introduced and delays were caused by landowners who insisted on detours to avoid their land or the spoiling of their view.

The Kennet & Avon Canal, near Great Bedwyn, Wiltshire

Tunnels, Locks and Aqueducts

The engineers most associated with canal building were James Brindley (whose work included the Bridgewater, Staffordshire and Worcestershire and the east and central Trent and Mersey), Thomas Telford (Shropshire Union and Caledonian), William Jessop (Grand Junction/Union), and John Rennie (Kennet and Avon and Lancaster). The

Locks on the Grand Union at Stockton, near Southam, Warwickshire

engineering feats accomplished were exceptional but the first phase of canal building avoided obstacles and followed contours, minimising the engineering and maintenance needed. This was understandable because there was little experience; continental canals inspired the Duke of Bridgewater but Brindley did not see them. Other problems were the lack of Ordnance Survey maps (the first were produced in 1801) and accurate geological surveys. Surveying techniques were also in need of refinement and it is a wonder that there are not more tunnels with kinks like Braunston Tunnel, the construction of which was also hampered by quicksands. In the second phase (1780-1820) canals went straight across difficult land. The Worcester and Birmingham Canal has 58 locks in 25.7km (16 miles), including the longest navigable flight of 30 locks in 3.2km (2 miles) at Tardebigge and several tunnels, including the 2,493m (2,726yd) long King's Norton Tunnel. The third phase coincided with railway building and involved improvements to the system such as the shortening of some of the contour loops on the Oxford Canal.

Aqueducts and tunnels are often impressive structures, particularly since the work was done largely by hand. The Barton aqueduct was the first to be built but Telford soon tried out a metal trough, supported on stone at Longdon-on-Tern on the Shrewsbury Canal, which avoided some of the problems encountered by Brindley. Telford followed with the Chirk and Pontcysyllte aqueducts on the Ellesmere Canal, near Llangollen, the latter, opened in 1805, being 305m (1,000ft) long. The Harecastle Tunnel on the Trent and Mersey Canal was the first long tunnel, 2.8km (1¾ miles) long, designed by Brindley and completed in 1777 after 11 years work. Telford built a parallel second tunnel in 1827 in only 3 years to allow a two-way system to operate. Brindley's tunnel is now closed. Standedge Tunnel on the now derelict Huddersfield Narrow Canal is the longest tunnel (nearly 5,029m, (5,500yd) and at times is 183m (600ft) below the surface) at the highest point of any artificial British waterway. The last tunnel was Netherton, built in 1858 with two towpaths and gas-lighting, as a by-pass for the narrow Dudley Tunnel. Sometimes cuttings were used and Woodseaves Cutting, on the Shropshire Union Canal, which is 1.6km (1 mile) long and 27m (90ft) deep, is one of several notable ones. Cuttings provide a good wildlife habitat because they often contain well-developed damp woodland with a variety of trees, shrubs, ferns and mosses.

Locks are also interesting features varying in shape and form; splay-sided, diagonal (to draw more water from a river into a canal as with the River Cherwell and the South Oxford Canal) and turf-sided (Kennet and Avon Canal). Most lock sides are faced with stone or bricks although some are rock-sided and others like Beeston Iron Lock (Shropshire Union Canal) were designed to offset particular problems such as quicksand. Locks were often grouped together into flights to save time, as on the Scottish Union Canal where all eleven locks were at Falkirk. Water supply is an important consideration when a narrow lock uses about 136,000ltrs (30,000gals) each time, a wide lock twice that figure and a Grand Union lock three times as much. Water was obtained from reservoirs at the summit level of the canal or at other strategic places, from rivers via steam pumps (at Leawood on the Cromford Canal and Crofton on the Kennet and Avon) or from rivers flowing into canals (from the Cherwell on the Oxford Canal and from the Trent for the Trent and Mersey Canal). Water shortages caused problems, particularly on the Oxford and the summit level of the Leeds and Liverpool Canal.Bridges may not at first seem exciting but there are large numbers carrying roads, access tracks, footpaths and the tow path over the canal. Bridge styles, often specific to a company, are distinctive for example those on the Shropshire Union Canal. The building material is often

A working lock on the Oxford Canal in Warwickshire

interesting, on the Oxford, south of Napton it is brick and nearer Oxford it is oolitic limestone. The different types of material used for bridges and lock sides and gates are reflected in the plants that have colonised them. Inclined planes are another engineering feat useful in hilly country where tub-boats (often with wheels) could be hauled up on tracks from one level to another. There were inclined planes on the Bude Canal (Cornwall/Devon), Tavistock (Devon) the Kidwelly and Llanelly, the Shrewsbury and others. The Anderton Lift was a late structure (1875) lifting boats in caissons 15m (50ft) from the River Weaver up to the Trent and Mersey Canal; originally steam-powered it was converted to electricity in 1908. The steam-powered Foxton Lift was even later, opening in 1900 and running for only 12 years.

Canals at Work

The cutting of canals provided jobs for many people; labourers, craftsmen and tradesmen, both local and from further afield. They included the cutters or navigators, miners for the tunnels, masons (for locks, tunnel linings and bridges), carpenters (for lock gates, wooden bridges and the frames for brick tunnels and bridges) brickmakers, quarrymen and provision merchants. Once a canal was built there was work for those on the boats and for toll keepers, lock-keepers and maintenance men (lengthmen). One hard job was legging, because few tunnels had towpaths. Boats were legged through by men, lying on their backs projecting on planks, who 'walked' on the roof or sides of the tunnel, while the horses went over the top. There was opposition from the poor who saw food disappearing into the towns, millers who lost water, turnpike operators who lost traffic and the owners of river craft unable to use canals. However, canals brought prosperity to towns, necessitating the building and running of warehouses. Stourport was one of the few towns in England which came into existence because of a canal. The Corporation of Bewdley, further upstream on the Severn, opposed the canal junction proposed first for their town and missed the benefits of the link between the Severn, Trent and Mersey via the Staffordshire and Worcestershire Canal.

From the start in 1757 with the St Helens Canal, until the rise of the railways after 1830 when the Liverpool and Manchester Railway opened, canals formed an important network of communication and trade, and with rivers, provided at their height approximately 6,400km (4,000 miles) of navigable waterways. The only exception were the Scottish Canals which were intended to cut out the difficult coastal journeys and, as short cuts, had none of the advantages of the inter-linked system south of the border. Even so the English system lacked some long distance links, for example to the north-east. The last canal to be built, the Manchester Ship Canal, completed in 1894, has been profitable, but in general the early canals were unable to compete with the railways and improved roads. Companies carried out improvements and lowered tolls but gradually canals became less economic, and closures began. As some canals became disused the diversity of plants increased and spread right across the channel as silt built up. The lack of management and disturbance made them more attractive to animals which took advantage of this change. However, there comes a time when the tall plants dominate the smaller ones, the water disappears and the wildlife interest decreases.

One of the oldest waterways in Britain, the Itchen Navigation, alongside the River Itchen near Brambridge in Hampshire. The water meadows to the right are extremely rich in plant and insect life

The railway companies accelerated the demise of some canals by buying the canal companies to eliminate opposition to the railway bills in parliament, to reduce competition from the canal carriers and in certain cases (for example the Croydon Canal) to use the line of the canal, including tunnels, for the railway. Canals bought by the railway promoters were supposed to be maintained in a viable state but this did not always happen and as the railways prospered, the canals went into decline, except during the two World Wars and a period in the 1930s. However, most of the major trunk routes have endured as a cross, linking the four major estuaries, Humber, Mersey, Thames and Severn, centred on Birmingham.

Most canals were nationalised in 1948, under the 1947

Transport Act, when the British Transport Commission was made responsible for them through the Docks and Inland Waterways Executive. For various reasons, the Thames, the Manchester Ship Canal, the Bridgewater Canal, the Yorkshire Ouse, the Exeter Canal, the Fenland Waterways and some disused canals were not nationalised. 3,200km (2,000 miles), about half the total of waterways, were nationalised and there were 6,000 working boats, some still privately owned. In 1948, with only a few exceptions like the Grand Union Canal, the canals remained much as they were when built at the end of the eighteenth century and commercial prospects for them in the mid-twentieth century were limited. By 1962, when, under the new Transport Act, the responsibility was passed to the independent British Waterways Board (BWB), the annual deficit was £1 million.

In 1965 the Board published *The Facts about the Waterways* in which it identified the few canals (mainly adjacent to estuaries) where commercial traffic might continue. It also recognised the potential of the rest of the system for recreation, pleasure cruising, angling and general amenity and its importance for water supply. The alternatives were to convert canals to drainage channels or to fill them in. These were expensive but non-recurring costs and had to be weighed against £340,000 yearly maintenance costs. The income in 1965 was approximately £70,000 from pleasure boating and £20,000 from fishing.

In the Transport Act (1968), the Board was given a new and positive role to provide for recreational needs on inland waterways. The Act defined three categories of waterways and their management:

a *Commercial Waters* 558km (347 miles) These are mainly river navigations, available and maintained for freight.

b *Cruising Waterways* 1,748km (1,086 miles) including 192km (119 miles) of river. These are principally used for cruising, fishing and other recreational purposes.

c *Remainder Waterways* 840km (522 miles) These were unnavigable for large boats in 1968. The Board has to maintain these at the state they were in in 1968; some, including the Brecon and Abergavenny Canal have been improved.

This leaves the disused canals which have become derelict for a variety of reasons, some closing as early as the mid-1800s, others as recently as 1964 (the Derby Canal). Some were short arms serving industrial premises or a city or town away from the main canal, as at Stafford. In others the cost of repairs was prohibitive, for example the Butterley Tunnel on the Cromford Canal in Derbyshire collapsed in 1900 isolating the northern section. The motorway building programme threatened several canals, but while the Midlands canals were protected, the Lancaster Canal only remains connected through culverts under the M6. Derelict canals include those which are being restored like the Kennet and Avon Canal, sections which have little open water and those where it is difficult to trace the original channel. Others, like the Derby Canal were deliberately filled in to form paths or roads.

Our waterway system still retains some of its incidental and important functions and although it is used less to provide power, the water from both canals and rivers is still very important for cooling purposes as indicated by the lines of power stations along major rivers such as the Trent.

A derelict lock basin on the Shropshire Union Canal near Abermule, Powys: now colonised by a variety of wetland plants as well as those found on drier banks

Rivers are managed, with flood banks and flood prevention schemes altering their courses. River straightening, tree removal and weed control practised along main rivers are designed to speed water away from towns and high grade agricultural land, which are liable to flooding.

Thus today there are nearly 4,800km (3,000 miles) of navigable inland waterways, both canals and rivers, compared with over 6,400km (4,000 miles) at the peak; there are 250 aqueducts, nearly fifty tunnels, over a thousand locks and thousands of bridges. The system is an important recreational resource particularly for angling and pleasure boating, which are most popular and also produce the most revenue. However canoeing on rivers and canals, power boating and water skiing, rowing on rivers such as the Trent, Thames and Severn and sailing on rivers and canal reservoirs are also in demand. Linked with the system are the unnavigable rivers which also form a network through lowland Britain. Together they form an important wildlife habitat which has developed alongside man's activities and it is this legacy, its variety and the problems associated with waterway usage that the rest of the book attempts to cover.

3 *Habitats*

The distribution of plants and animals along canals and rivers may sometimes seem haphazard, but there is a combination of circumstances that means that a particular species is present, or perhaps more importantly, absent. Although animals have other requirements which are often quite precise, (for example oxygen levels) the basic need, either directly or indirectly, is the presence of plants for food and also for shelter and breeding sites. The more diverse the habitats and vegetation the greater the variety of small animals will be. Certain plants or animals become numerous and may create problems where physical or chemical factors are less diverse, usually after the balance of nature has been upset by man.

The factors affecting plant and animal life fall into three groups, all affecting the presence or absence of a species and its degree of success. *Physical features,* such as flow, stream bed type, width and depth of channel, light and temperature, create diversity in the habitat, increasing the opportunities for wildlife. *Chemical factors,* such as dissolved gases and nutrients may be affected by the current, the temperature or the type of rock. Both the chemical and physical qualities may be altered by *man's activities,* for instance, management to prevent flooding, as well as pollution and recreational use of the waterways (covered in chapter 12). Not only do these three sets of factors interact but each varies in its effects at different times of the year.

Flow

The upland stages of rivers are fast flowing, because the gradient is steep and the rainfall high; lower down the valley bottoms are flatter and the current is slower. In canals the use of locks creates a flow, but there is usually more water than necessary and it flows over the side weirs near locks. These weirs often contain plants and animals, for example the moss *Fontinalis,* which can tolerate the higher rate of flow. In contrast the flow is very slow in some of the long pounds where locks are twenty or more miles (32km) apart.

The speed of the water in a stream varies greatly even in short stretches, where the bottom and the gradient may be very different. In both rivers and canals the water moves fastest in the middle or main channel, on the outside of bends and on or just below the water surface. Plants which grow in the slower water at the edges slow it down even more. Other obstructions, such as rocks and bridges, slow the flow but increase the turbulence or roughness of the water. Rough, slow flowing water is more important than a smooth flow in determining which plants and animals can live in a particular stretch. The current and the turbulence also affect the turbidity or cloudiness, caused by silt and mud in the water.

Generally, water courses with similar flows have similar vegetation, whether they are in the north, south, east or west. Fast flowing rivers usually have a high proportion of mosses and liverworts and no flowering plants. Only lower down where the flow is slower does the variety of plants increase. Canals may be important, for example in

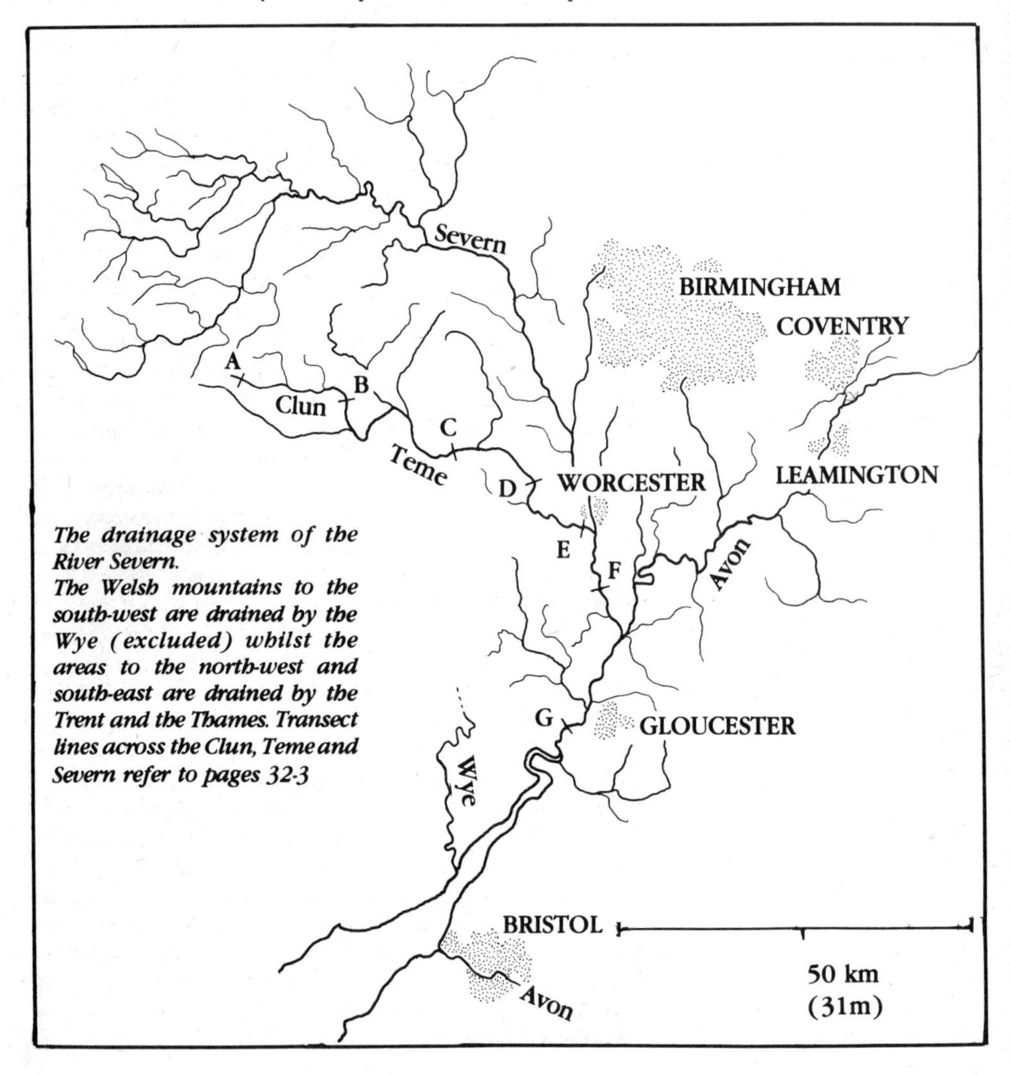

The drainage system of the River Severn. ***The Welsh mountains to the south-west are drained by the Wye (excluded) whilst the areas to the north-west and south-east are drained by the Trent and the Thames. Transect lines across the Clun, Teme and Severn refer to pages 32-3***

Scotland, in providing a slow flowing water body in contrast to the fast rivers which dominate the area. Anything that slows the water down helps to add diversity; the River Weaver in Cheshire is poor in species except at Winsford Flash where the drop in water speed and the subsequent changes in other conditions has encouraged a range of flowering plants in contrast to the mosses and liverworts which are dominant elsewhere.

The most obvious problem that flowing water presents to plants and animals is the difficulty of staying in one place. Some plants overcome this by developing strong roots or holdfasts to anchor themselves to rocks in fast-flowing streams where there is little mud or silt. The mosses and crowfoots, *Ranunculus* species, which grow in fast streams live horizontally in the water because any vertically growing plants would be continually battered. The flowers of the crowfoots grow above the water in summer when the flow is reduced. In slower water, emergent plants are firmly anchored in the mud or silt by roots or rhizomes and free-floating plants such as algae and duckweeds can also survive. Uprooting is obviously more damaging than losing or injuring stems and leaves which can be replaced or repaired. In the upper reaches oxygen and nutrient levels are usually high and growth can be rapid, but spring floods on lowland stretches can hit plants at a bad time before the roots which died back in the winter have regrown.

Another adaptation by plants is the rearrangement of the inside of the stems and a reduction of the stiffening materials, so that they are tough enough to withstand the pull of the water but flexible as well. Leaves are often thin and flexible and plants like arrowhead have strap-like underwater leaves, broader floating leaves and aerial ones.

The fish that live in the faster flowing streams such as bullhead, *Cottus gobio,* and stone loach, *Noemacheilus barbatulus,* are flattened on the lower (ventral) side and are streamlined, so that if they lie head pointing upstream the water flows over them. The salmonid fish, trout, *Salmo trutta,* and salmon, *Salmo salar,* are streamlined but are also very powerful swimmers. They are big fish and need a good depth of water to move upstream to their spawning grounds. Unfortunately, in some places water abstraction has evened out the flow and they can not get upstream. Invertebrates in faster streams have flat bodies and often live under or among rocks, for example flatworms and the water limpet, *Ancylastrum fluviatile.* Some, like black fly larvae, *Simulium* species, have hooks, leeches have suckers and others tie themselves down with a bed of silky threads. One of the additional benefits of cementing small pebbles into a protective case by some caddis fly larvae is that they are weighted down. Remaining stationary complicates feeding and the animal has to wait for food to

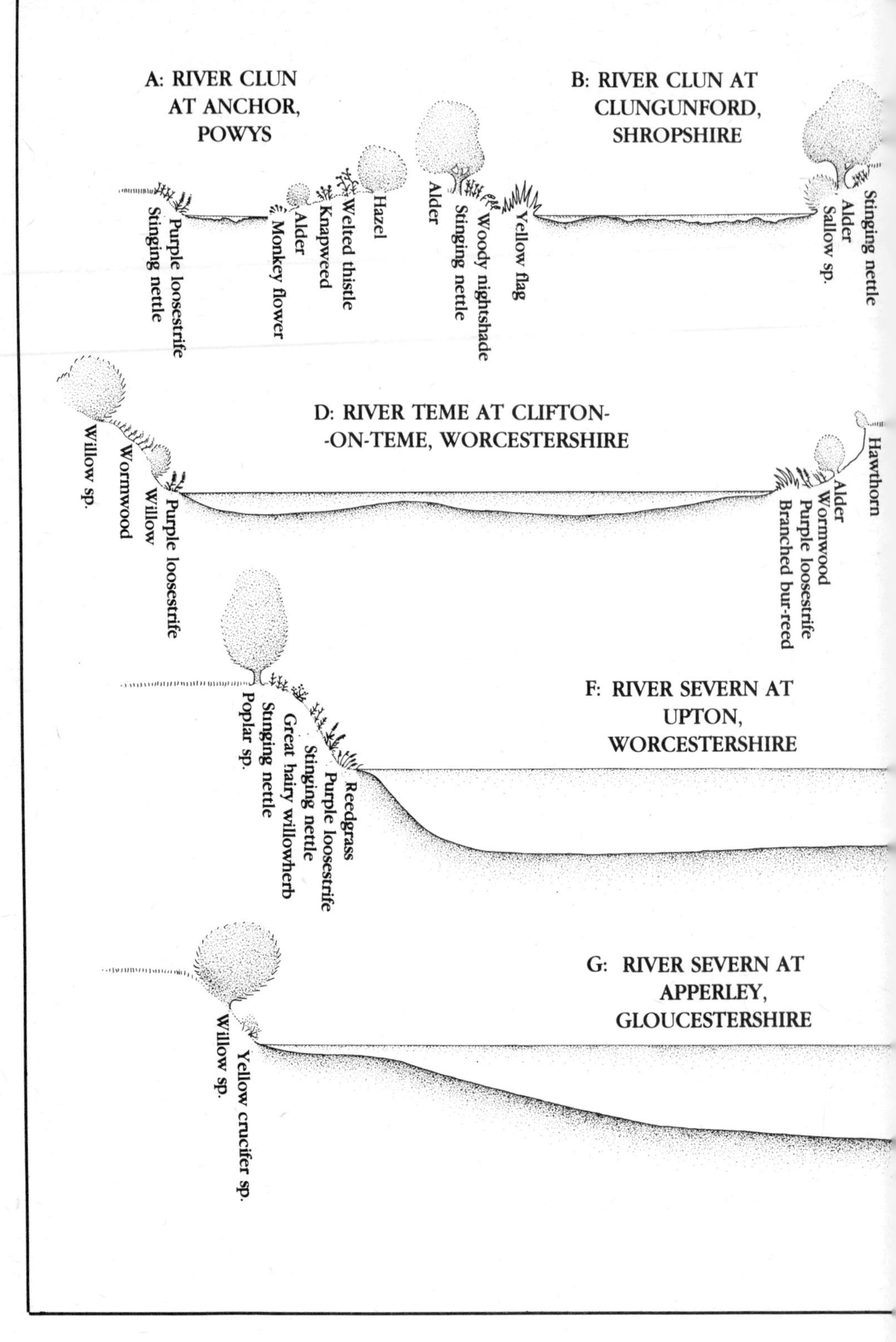

A: RIVER CLUN AT ANCHOR, POWYS
Stinging nettle
Purple loosestrife
Monkey flower
Alder
Knapweed
Welted thistle
Hazel
B: RIVER CLUN AT CLUNGUNFORD, SHROPSHIRE
Alder
Stinging nettle
Woody nightshade
Yellow flag
Sallow sp.
Alder
Stinging nettle
D: RIVER TEME AT CLIFTON-ON-TEME, WORCESTERSHIRE
Willow sp.
Wormwood
Willow
Purple loosestrife
Branched bur-reed
Purple loosestrife
Wormwood
Alder
Hawthorn
F: RIVER SEVERN AT UPTON, WORCESTERSHIRE
Poplar sp.
Stinging nettle
Great hairy willowherb
Stinging nettle
Purple loosestrife
Reedgrass
G: RIVER SEVERN AT APPERLEY, GLOUCESTERSHIRE
Willow sp.
Yellow crucifer sp.

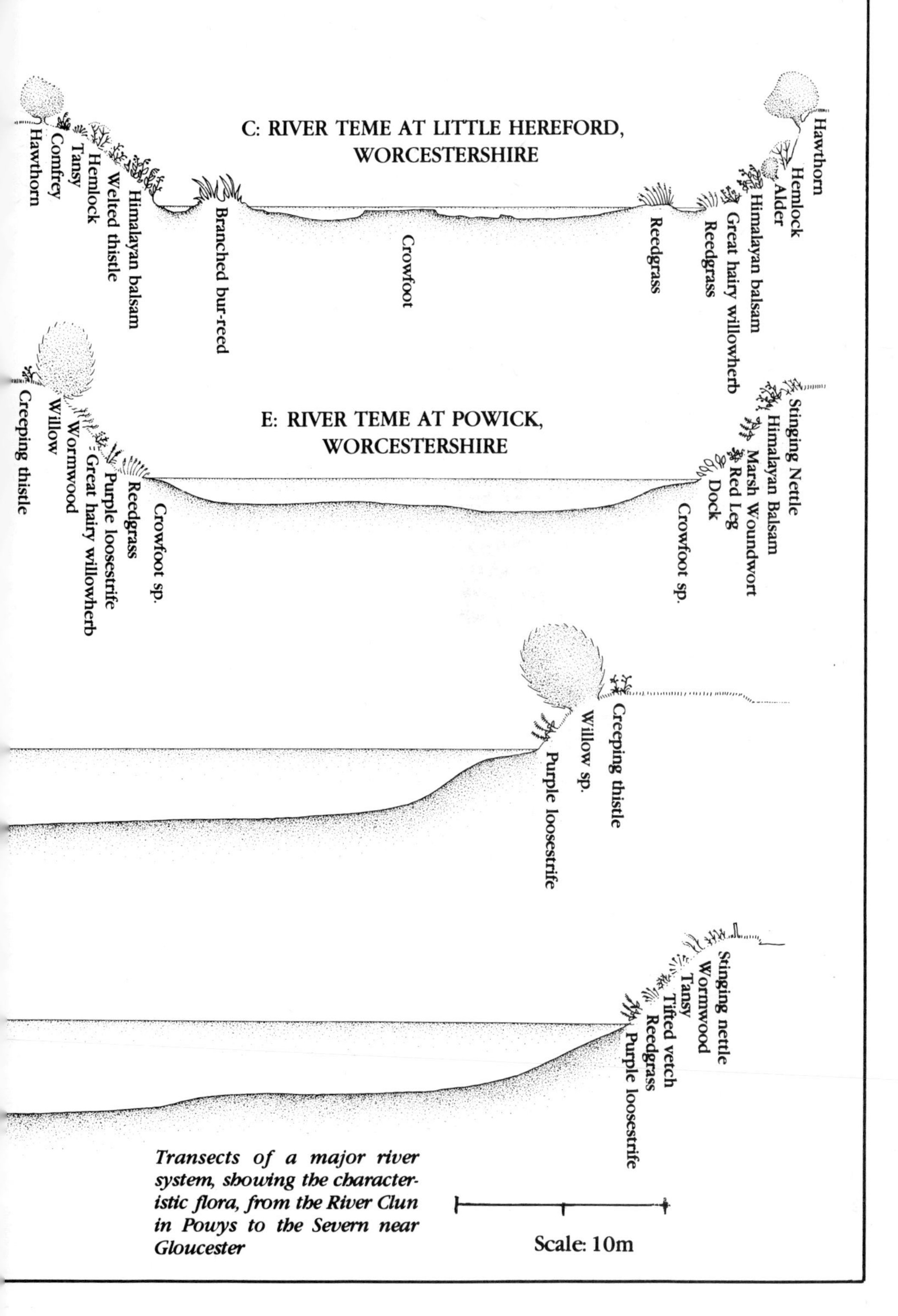

Transects of a major river system, showing the characteristic flora, from the River Clun in Powys to the Severn near Gloucester

come to it. The net-spinning caddis larvae filter water through the net and trap food particles.

Sticklebacks and other lowland fish are not ventrally flattened and the invertebrates have fewer specialised adaptations, because the flow is slower and plants provide shelter and food.

Substrate

One of the greatest differences between canals and rivers is the relative uniformity of the clay puddle lining of canals compared with the changes from rock, through pebbles and sand to silt as the river goes from source to sea. Canals may have sections cut through rock and others faced with stone, but the slow flow usually results in silt and organic matter settling on the bottom. In the upper reaches of rivers, rocks dominate and the pebbles, sand and silt eroded there are deposited further downstream. The type of rock affects the stream bed; chalk streams have little sediment, because it dissolves. Silting is greater from sandstones and granite, which produce a variety of particles and in clay streams which are turbid. The leaves from overhanging trees each autumn may make a very organic acid mud where only anaerobic animals can survive. The size of the particles and the stability of the stream bed are governed by the flow as well as by the rock type. Fast flowing water knocks pebbles together, increasing the likelihood of breaking; where the flow is slow deposited mud is relatively stable. On the outside of a bend erosion occurs, often cutting steep 'cliffs', but on the inside where the flow is least there is a build up of sand and silt. In any river, plants and animals may be responding to flow, substrate or both. Mosses grow on rocks because if they grew on mud they would be constantly undermined and covered with silt; rocks are dominant in places where silt is not deposited because of the fast flow. None of the flowering plants grow on rocks, they root in silt and mud, which are only deposited where the flow is slower. Plants may get broken but some like the milfoils, *Myriophyllum* species, can regrow from fragments. Horned pondweed, *Zannichellia palustris*, can grow through silt as it is deposited so that the roots are always near the surface, but it is very vulnerable when the silt bank is flushed away in floods.

Sand and silt carried in water can be very abrasive and plants and animals (for instance lowland stonefly and mayfly larvae) are hairy to reduce the damage. Some plants have aerial leaves, but even they have to be able to withstand damage while they grow through the water. Some mayfly larvae burrow to avoid the silt and others, for example *Caenis* species, are able to shut off the gills on one side so that the silt-free side breathes and avoids clogging. Silting can reduce the numbers of snails, midge larvae, water lice and other invertebrates that are important as food for larger animals, and cloudiness can severely limit the ability of fish to hunt for prey.

Channel Width and Depth

Width is not as important as depth in affecting the plants and animals colonising an area, although in narrow channels the flow is concentrated and there is less scope for bends, shallows and other variations which provide more habitats. Depth affects the amount of light reaching

Reedgrass,* Phalaris arundinacea, *is often found on gravel spits in rivers

the bottom, the stability of the bed, temperature and the level of dissolved gases, all of which are also affected by the flow.

The shallower a stream is, the more light can penetrate the water (unless the silt content is high), but on the other hand it may be more likely to dry out in summer. In disused canals and man-made dykes the depth can be reduced gradually by silting and a build-up of organic matter until it is effectively filled up with vegetation and no open water remains. In rivers this is unlikely to happen unless the flow is very slow because floods will erode the vegetation and silt every so often. Some plants cannot grow under water; reed-grass, *Phalaris arundinacea,* needs to be above water all the time but common reed, *Phragmites australis,* will grow well in water up to 1.3-1.6m (4 or 5ft) deep. Animals, like plants, space themselves vertically according to the habitats and the water, but because they are not limited by light they can go much deeper as exploration of the ocean deeps has shown. It is usually light, temperature and other factors associated with the depth which are critical, not the depth itself. Canals usually have a regular depth profile although it varied according to the original engineers, whereas rivers have shallows and deep pools within very short distances.

Light

Water reflects light at the surface and then absorbs the red part of the light spectrum that does get through, causing everything below 3m (10ft) to have a bluish tinge. Unfortunately, red light is important for photosynthesis, the process by which plants manufacture food, and therefore they can only grow in shallow water. Some plants avoid the problem by having floating or aerial leaves, but the roots have to store a large amount of food in spring so that the leaves grow through the poorly lit zone. The amount of light is also affected by the turbidity and discolouration such as ochre, peat or pollution.

Plants arrange themselves in water according to the light; some pondweeds, *Potamogeton* species, manage with less light than others. Other plants have leaves at different levels to maximise the use of the light. Bur-reeds *Sparganium* species, and other emergent plants will grow in shade but others such as arrowhead, *Sagittaria sagittifolia,* and brooklime, *Veronica beccabunga* tend to flourish in the open.

Flow and other factors may mean that light is not as fully used as it may be in a lake because free-floating phytoplankton (minute plants) are less common and part of the available area is unoccupied. Canals, being more uniform, fairly shallow and slower flowing are more efficient but boats churn the silt up increasing turbidity and counteracting any beneficial characteristics. Animals depend on plants, but are less directly dependent on light, unless it enables predatory fish to hunt.

Temperature and Oxygen

These factors, one physical the other chemical are linked and are best dealt with together, because oxygen is relatively insoluble in water and the solubility decreases as the water gets warmer.

The temperature of a river will depend on the stage the river has reached, its turbulence and depth, shading and the time of day. The origin of the inflowing water is also important because water moving through the ground (ground water) is generally colder than water flowing over the surface. The temperature of running water follows that of air, closely and quickly. Upland stretches are in colder parts of the country, but the turbulence which is more frequently met there increases the contact surface area and mixes the air and water, hindering freezing in winter but also allowing large rises in summer. Shade prevents the sun warming the water and depth in slower flowing water reduces the temperature. In canals the temperature pattern is more like that of lakes, where there is a wider daily and yearly range compared with rivers; the coldest water is at the surface in winter and at the bottom in summer. The surface water of canals often freezes and this reduces the further lowering of temperature. In contrast, rivers usually remain unfrozen, although in exceptionally cold winters even the Thames has frozen and fairs have been held in the middle of the channel!

Water from canals and rivers is commonly used for cooling in factories and power stations. While the resultant heated water, which is returned to the river or canal, is a form of pollution its main effects are the same as natural temperature changes, although exaggerated in scale. Thermal pollution is very noticeable on cold days when a mist rises from the river surface downstream.

A look at the mean temperature figures at different places on river systems, for example the River Trent in Staffordshire, Derbyshire and Nottinghamshire, can be related to power stations and to sewage treatment works, even where major tributaries enter and increase the flow. The tributaries are often colder than the main river; for instance the Dove, near the confluence has a mean temperature of 10.2°C, for the Mease the figure is 10.4°C and the Trent is already 11.0°C at that point. While these are only mean figures, hiding daily and weekly variations the figures for Drakelow Power Station, near Burton-on-Trent show the rises that are possible. The temperatures of intake water recorded on one day in summer and in winter are 18.2°C and 6.6°C respectively. The heated water returned to the river at the same time is 27.4°C (summer) and 20.4°C (winter), that is an increase of 9.2°C and 13.8°C respectively. What is more important is that 1.6km downstream the temperatures are still high, 22.4°C and 12.4°C respectively. Casual recordings of the Trent and Mersey Canal at Stenson near Derby and the River Trent (downstream of Willington Power Station) less than 1.5km away, reveal differences of around 8°C in summer. However, because of the way the water falls down the cooling towers in power stations, it is more oxygenated when it returns and this counteracts some of the oxygen loss likely to be caused by the increase in temperature. Other cooling processes may not result in such agitation and subsequent oxygenation. Canals have been heated artificially by cooling water from mills in Lancashire and Yorkshire and exotic plants became established including *Najas graminea,* a narrow leaved pondweed and also tape grass, *Vallisneria spiralis,* which has also colonised rivers in similar circumstances. Various tropical worms and fish, for example the guppy, have also colonised warm water areas.

Power stations like this one at Willington on the River Trent use water for cooling

Temperature and Flow of the River Trent

Distance between sampling points	Sampling Points	Power Station and distance above sampling point	Mean temp 1974-80 °C	Mean flow Ml/d 1979-80
	Great Haywood		11	c500
20km				
	Yoxall	Rugeley 8km	11.2	1362
14km				
	Walton below Tame confluence		11.1	3558
16km				
	Willington	Drakelow 11km	12	4988
21km				
	Shardlow	Willington 19km Castle Donington 2km	15.6	5098
3km				
	Sawley below Derwent confluence		13.7	6981
17km				
	Trent Bridge, Nottingham	Ratcliffe on Soar (indirect cooling) 14km	13.9	9063
20km				
	Gunthorpe		13.6	9245
17km				
	Kelham	Staythorpe	15.4	9469

Higher temperatures will generally speed up plant growth although the oxygen supply may deteriorate and reverse the trend. All invertebrates are cold blooded and cannot regulate their body temperatures, which follow that of their surroundings. Therefore they occur in greater numbers in summer and their life cycles are geared to maximum growth and reproduction then, passing the winter as resistant stages such as eggs or pupae. Some fish have high metabolic rates and can be active and feed in cold water but die in temperatures over 20°C, whereas others with a lower rate (such as perch) may be very sluggish in cold water but increase their activity up to 30°C. However, oxygen is also a factor because fish occurring in colder upland water, like brown trout, grayling and bullhead have a higher oxygen requirement than carp, tench and bream. Roach, pike and perch are intermediate and occur in the middle reaches. The upper reaches have colder water and a lot of turbulence, both of which increase oxygen absorption. Lower down the temperature is higher, the water is smoother, there is breakdown of organic matter and a higher plant and animal population, all combining to lower the oxygen level. Canals are slower moving and generally have less oxygen, and those that are disused often become stagnant. Turbulence is artificially created in both canals and rivers by locks, weirs and boat propellors. Oxygen is released by plants during photosynthesis but they still need it for respiration, and at night when photosynthesis stops in the absence of light, oxygen levels may fall dramatically. Plants have to absorb oxygen where they are and the water milfoils and submerged pondweeds have feathery leaves to increase the surface area.

Animals have various adaptations for obtaining oxygen; some beetles take in a supply from the surface and store it under their wing cases (elytra), pulmonate snails rise to the surface to get a supply before returning to their food below. Mosquito larvae have abdominal breathing tubes which they use at the surface; in the next stage, the pupa, the tubes have shifted to the head! Generally, small animals, such as worms, leeches and flatworms do not need much oxygen and with a large surface area they absorb enough through their skins. Haemoglobin, which has a great affinity for oxygen is not usually found in invertebrates but it occurs in bloodworms (midge larvae), the snail, *Planorbis* species, and tubifex worms, enabling them to live in virtually anaerobic conditions. Other invertebrates such as mayfly and stonefly larvae have developed gills with large surface areas for breathing. Amphibian tadpoles have two pairs of gills when they hatch and produce another pair as they grow in size and become more active.

Where pollution is high, particularly sewage and farm effluent, a lot of oxygen is used up in the decomposition of organic matter. In heavily polluted stretches of waterways the tubifex (sludge) worms, and rat-tailed maggot (hoverfly larva) are most numerous. Water lice and bloodworms can survive moderate pollution and in cleaner water caddisfly larvae and freshwater shrimps are replaced by stonefly and mayfly larvae only where there is no pollution. In derelict canals, once the oxygen levels drop, only bloodworms survive in large numbers; dragonfly and caddisfly larvae disappear and snails are less frequent. Gradually the fish die and as the vegetation builds up — only moorhens and water voles increase.

Nutrients

More chemical compounds will dissolve in water than in any other natural liquid, but the amount and types will depend on the geology and the amount of dilution. Rivers flowing from igneous rocks are low in salts with only small amounts of calcium, magnesium, sodium and potassium as bicarbonates, sulphates and chlorides. The silica level is often high and this is important for diatoms (minute free floating organisms), which use it to construct their shells. The supply may be exhausted later in the season causing a population crash. Sedimentary rocks produce greater amounts of dissolved materials especially calcium, which is important for the shell-building molluscs. Slightly acid rainwater, containing sulphur dioxide from pollution or natural carbon dioxide will act on rocks and eventually dissolve them. Naturally occuring trace elements such as boron and molybdenum are also important. Some nutrients, such as phosphorus, which are in short supply may be further depleted when they are locked into the bodies of plants and animals or as insoluble compounds in the sediment.

Generally, there is a natural process of enrichment, eutrophication, from source to the sea. Upper reaches are nutrient-poor (oligotrophic) but as more are added to the water from run-off, solution of the bed rock, deposition of silt, and the breakdown of plants and animals, it becomes richer. Canals do not exhibit the same changes; they do flow in one direction even though the bed is uniform but the levels of nutrients only vary widely when agricultural run off and other non-natural sources are added. However there can be gradations which show in derelict canals where the flow and absence of traffic result in little mixing or dilution. In the Cromford Canal in Derbyshire there are some areas where inflows are from limestone and others where acid gritstone and shales occur and the plant communities are different.

It is probable that nutrients are rarely limiting for plants and although they may stop growing they do not die. Many aquatic plants absorb nutrients from the water through stems and leaves and also have roots which extract them from the silt. One example of specialised adaptations is those plants and animals which are able to live at the river's mouth or estuary, where fresh and salt water meet. They have to cope with a range of salt concentrations, turbidity, changes in temperature (in summer fresh water is warmer than sea water and the reverse is true in winter), water level and current, all of which change markedly throughout the tide cycle. Consequently relatively few plants and animals have colonised estuaries, compared with other aquatic habitats. The soft muds encourage burrowing worms, but discourage many higher plants. However there are some plants which normally grow on the shore which find suitable conditions inland where natural salt occurs. For example, sea spurrey, *Spergularia marina,* wild celery, *Apium graveolens,* orache, *Atriplex glabriuscula,* a rush, *Juncus gerardii,* and a sedge, *Carex distans,* all occur on the Droitwich Canal.

Canals and the Spread of Plants and Animals

Canals have introduced one new factor into plant and animal distribution. They cut across hills carrying plants and animals over watersheds such as the Pennines. Even though the sources of two rivers may be less than half a mile apart, aquatic organisms cannot bridge the gap. Boats

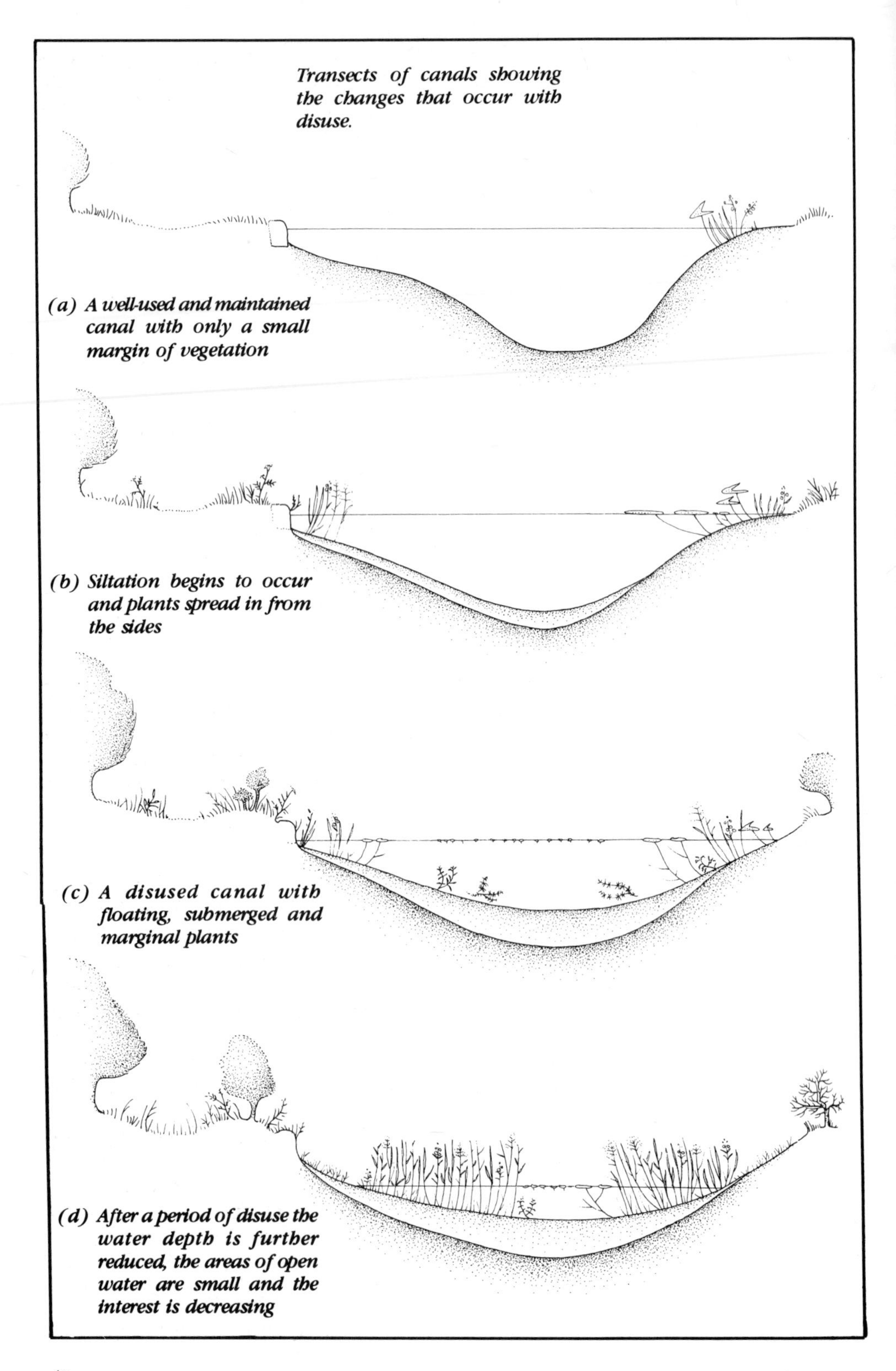
Transects of canals showing the changes that occur with disuse.
(a) A well-used and maintained canal with only a small margin of vegetation
(b) Siltation begins to occur and plants spread in from the sides
(c) A disused canal with floating, submerged and marginal plants
(d) After a period of disuse the water depth is further reduced, the areas of open water are small and the interest is decreasing

might travel down a river, out to sea and up another river but the freshwater organisms are unlikely to survive the salt-water, even if they manage to stick to the boat. Before canals they relied on being carried accidentally, for instance by ducks; once the short cut was there the effects must have been dramatic.

One of the best documented examples is Canadian pondweed, *Elodea canadensis,* an introduction, first recorded in England in 1846 in a canal reservoir, near Foxton Locks, on the Grand Union Canal, near Market Harborough. The canal was only thirty years old then but according to the lock-keeper the plant had been there for twenty years. Foxton is fairly central in the network and the pondweed spreads easily by fragments which develop into new plants.

Native plants and animals spread well too, although they are not documented; certainly canals are well colonised now. Two plants of high acid lakes, floating water plantain, *Luronium natans,* and the starwort, *Callitriche hermaphroditica* moved into the lowlands via canals. Some of the pondweeds, *Potamogeton* species have met up with relatives they had not encountered before. Several have hybridised, and because the resulting plants were new they caused confusion amongst botanists. Slender rush, *Juncus tenuis,* another introduction, has found towpaths to its liking, progressively colonising them, for example those on the Basingstoke Canal in Surrey. There are similar success stories in the animal kingdom where species have colonised new areas by moving along the canals. One example is the invertebrate, *Corophium curvispinum* mentioned on page 74.

Habitats

The various physical and chemical qualities of rivers and canals combine and interact to provide habitats for plants and animals. Canals have no natural couterpart and a look at the bed reveals the first major difference; canals have most of their habitat diversity spread across the channel, with deep water in the middle and shallows on either side. Except in places like Tixall Wide on the Staffordshire and Worcestershire Canal and subsidence areas, most canals are narrow and one stretch is basically like another. The puddled clay base which may be 45-100cm thick evens out the substrate and the deposited silt is a good growing medium. In contrast the bed of a river is very variable, both across the channel and along its length. It may be narrow or wide and there may be pools, shallows, waterfalls, meanders, acute or gentle bends and the bed can be solid rock, gravel, sand or mud all within a mile. The upper and middle reaches have a more varied profile as pages 32-3 show with islands, spits and shallows whereas the lower reaches are deeper and exposed areas, usually the sides, are muddy. Sand and gravel spits may be colonised by certain plants, like reed-grass, *Phalaris arundinacea.* Common sandpipers, or little ringed plovers may breed on them while wagtails may use such places as a base for foraging. Protruding rocks are good vantage points for birds such as the dipper, a species not confined to upland rivers. Shallows are important spawning areas for fish like grayling, where the gravel encourages aeration. Pools are the favourite haunt of adult salmon but they need shallows for spawning and find their food upstream of rough water (riffles). Coarse fish do not need so much variety in the bed, but they do require vegetation for protection. Once there are pools, shallows and spits, there is also an increase in flowering plants and the mosses and liverworts, which may form more than half of all the plants in upland stretches, become less dominant.

A typical canal profile has the main channel slightly off-centre, a gradual shallow on one side and a deeper margin on the other. Gravelly patches can occur where wash from boats has scoured the bank above the puddled clay lining. Thus canal and river margins are similar and some rivers, like the Worcestershire Avon, have a marginal shelf just like a canal but wider, (90cm or 2½ft). Bulrush, *Schoenoplectus lacustris,* which grows in water, and raft-building plants like reed, *Phragmites australis* help to make up the rich marginal community. The dense emergent vegetation is important for nesting moorhens, coot, little grebe, mute swan, mallard and the shy water rail; reeds shelter reed warblers. Canals are probably safer nesting territory than rivers, which may flood in spring, swamping the nests although wash from boats is also hazardous. The lack of floods in canals allows silt to accumulate. In a well-used, dredged canal, vegetation is confined to the edges, but with less use and less maintenance siltation and vegetation increase until in derelict canals the amount of open water is very small, (see page 42). The variety of aquatic plants and animals disappears as dominants such as bur-reed, *Sparganium erectum,* take over.

New river works may eliminate some of the marginal habitats but the sloping stone-facing in some makes ideal niches for small plants as well as wagtail 'parade grounds'. Steel piling on canals often cuts off an area of still water protected from wash where fragile plants can survive. Vertical earthbanks, which are more common on rivers than canals, give scope for nesting kingfishers, sand martins and dippers.

The marginal and bank vegetation is very important for fish and invertebrates providing shelter and shade. Both banks of a river may be grazed or cultivated leaving them unshaded, except perhaps for the occasional tree; the wooded stretches that do exist are a reminder of the forests and swamps through which post-glacial rivers flowed. The

Willows beside canals are often pollarded and then cut again on the canal side. (The Trent and Mersey in January)

mixture of shade, light and dappling is important in creating variety. Old oxbows ('mortlakes') are often rich in plants and animals including rarer ones which have been eliminated from the main channel by pollution or management since the separation.

In most canals the two banks are managed differently. The offside bank may be wooded, cultivated or grazed like a river bank; in summer, cattle may spend time in shallow water cooling off, drinking, eating the marginal plants, trampling the habitat and creating bare mud. The towpath side used to be grazed by the towing horses when boats had moored for the night, but it is cut now and is rough marshy or dry grassland with a hedge at the boundary. Rivers and canals which pass through villages, towns and cities are often landscaped as parks (the Tame) or stepped embankments (the Trent) for amenity and safety which removes the variety of habitats. Canals may also have small marshy areas adjoining the bank just like rivers.

Trees, copses and woods are important; shade creates a more complex habitat. Many different types of algae will coat stones in a shaded stream and mosses, ferns and liverworts occur on the lower banks too. The damp-loving woodland plants are also there to offer more variety than agricultural grass or crops. Trees are important for animals; alder, *Alnus glutinosa,* attracts siskins and redpolls which eat the seeds on the trees, while mallard and teal eat them on the water. Some woodland birds, like lesser spotted woodpecker and pied flycatcher are often found in waterside woods.

The traditional management of willows, *Salix* species, which is called pollarding, provides additional habitats. In order to protect the new growth from being eaten or browsed by cattle the trees are cut off at 2-3m (6-10ft) above the ground so that they grow from this new 'crown'. The branches are used for making hurdles and baskets. Repeated pollarding often results in a very thick trunk and large crown which may have pools of water in it. These pools have their own collection of insect larvae and beetles. Dry pollard crowns are safe roosting places for many birds away from predators like foxes. Many pollards eventually become hollow and may provide a roosting place for owls. Trees are also important in determining which lengths of river bank provide suitable resting and breeding places for otters. Alders and willows have very fibrous roots which bind the bank together. The more open roots of ash, *Fraxinus excelsior* and the introduced sycamore, *Acer pseudoplatanus,* allow the bank to be washed away to form a 'holt'.

Scrub is another important habitat generally and one that often occurs by canals and less so by rivers. Mixed willow-scrub is rich in insects and its much-branched habit provides shelter and nesting sites for sedge warbler,

An otter territory on a Northumberland river: a good supply of fish, little disturbance and plenty of tangled cover on the bank are the main requirements

Acrocephalus schoenobaenus, reed bunting, *Emberiza schoeniculus* and other summer visitors. Part of the scrub element occurs in canalside hedges which are a feature not commonly found by rivers. Hedges are important because they shelter the water and provide another habitat, although many of the original ones have been replaced by fences.

Truly artificial structures, bridges, locks and weirs also support living things. The gates and sides of locks have a range of mosses and liverworts that tolerate immersion and above water level ferns, willowherbs, *Epilobium* species, skull-cap, *Scutellaria galericulata,* and gipsy-wort, *Lycopus europaeus* find a precarious foothold. Weirs provide fast water habitats. Canal reservoirs are often important wildfowl areas in winter and one or two contain interesting plants. However, nesting birds and the marginal vegetation may be left high and dry as the water is used in summer.

The major habitats all contain micro-habitats which are equally important; different sized gravel produces a network of turbulence, silting and aeration within a small area, with niches for a variety of organisms. Where one plant has become established, the water slows down and others take advantage of the deposited silt and nutrients. The roots may shelter invertebrates and they burrow into large stems and leaves. A clump of long-leaved water crowfoot, *Ranunculus fluitans,* may harbour snails which are eating the middle away. Hornwort, *Ceratophyllum demersum,* with its stiff, forked leaves protects small animals better than others with softer leaves. Other flowering plants may not be important for some invertebrates, except as a surface on which their food, algae, grows.

Thus each stretch of river or canal or marsh has a community of plants and animals which, to a greater or lesser degree, rely on other members for their survival. The idea of a food web is familiar to most people. At its simplest carbon dioxide activated by sunlight in photosynthesis produces growth in plants which are eaten by herbivorous animals; these in turn provide food for carnivores, such as dragonfly larvae which are caught by larger animals. Plants and animals die and are cleared up by scavengers, detritus feeders and bacteria, releasing some of the nutrients so that plant growth is possible again. As it is continuous the whole process does not grind to a halt until the scavengers have finished, although there are times when some nutrient or vital element, for instance silica, is locked up in animals or plants and limits the growth of a group of organisms. Rivers and to a lesser extent canals are losing nutrients as parts of plants, dead and living animals are carried downstream, and adult dragonflies, mayflies and amphibians leave, impoverishing the system.

Even apparently inhospitable sites such as lock gates are colonised by such plants as gipsywort, **Lycopus europaeus**

4 Aquatic and Emergent Plants

Life in rivers, streams, ditches and canals is not always easy. The difficulties of coping with flowing water, of obtaining nutrients and reproducing have been overcome in a variety of ways by colonising plants.

Living in Water

Most aquatic plants have far less strengthening (lignified) tissue than land plants because the water supports them in a way that air does not. However, the channels for carrying water and food within the plant have moved from the outside of the stem (where they are in land plants) to form a central core which withstands the pull of the water. In addition, air spaces (aerenchyma) between star-shaped cells have evolved to give the plant buoyancy and act as a store for oxygen. Roots may be only holdfasts, although in some plants they are able to get nutrients from the soil or silt just as in land plants.

One problem that faces land plants is water loss, but aquatic plants do not wilt because water is never in short supply and is absorbed through the surface. The 'current' which carries nutrients around the plant is created when water is lost (transpired) through pores and in land plants these are often on the underside of leaves where they are less exposed to the drying air. By contrast the pores or stomata are usually on the upperside of floating leaves so that they can transpire, although in frog-bit, *Hydrocharis morsus-ranae,* and broad-leaved pondweed, *Potamogeton natans,* they remain on the underside. Aquatic plants have diaphragms at stem nodes like bulkheads in submarines to prevent flooding of the whole plant if damage occurs. Nutrients and gases (oxygen and carbon dioxide) are taken in all over the plant and so finely divided or broad flat leaves are commonly developed to increase the surface area; they are often then without a hardened skin or epidermis which would slow down absorption. Chlorophyll, the green pigment, lies near the surface and absorbs carbon dioxide for photosynthesis, leaving crusty deposits of calcium carbonate on the leaves. Some plants, for instance, arrowhead, *Sagittaria sagittifolia,* take no

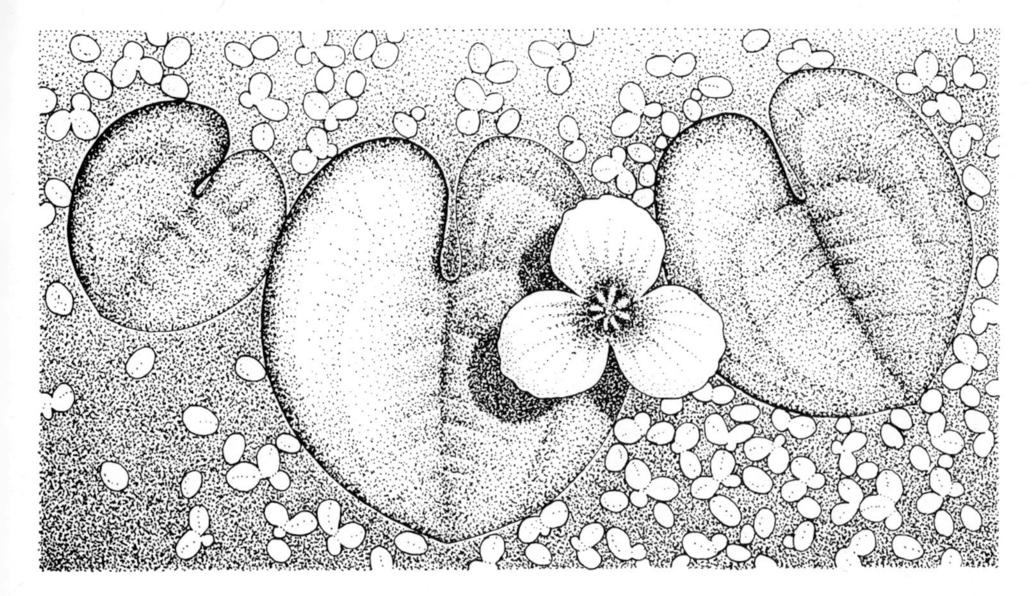

The heart-shaped leaf and three-petalled flower of frog-bit, **Hydrocharis morsus-ranae,** ***with fronds of duckweed,*** **Lemna minor**

chances and have aerial, floating and submerged leaves, all differently shaped, so that they can get what they need from the air, water or the silt in which they are rooted.

Reproduction is difficult in water and although some submerged plants like hornworts, *Ceratophyllum* species, pollinate normally, many have tiny self-pollinating flowers and others raise their flowers above the water. Some seeds are light with inflated stems or slime casings to make them float better. Others stick or hook onto animals, and some just drop to the bottom. Vegetative reproduction, the growth of a new plant from a fragment or special bud, is very important for some plants where flowering is spasmodic. Only one sex of Canadian pondweed, *Elodea canadensis*, was introduced at first and even now that both are present the female plants rarely flower; its spread by vegetative means alone was, therefore, very impressive.

Overwintering

The way that plants survive the winter is important; some, like the hornworts, produce winter buds which have fleshy food-packed scales and which sink in late summer, ready to produce new plants in the following spring. In duckweeds, *Lemna* species, the old plants become heavier by absorbing calcium in autumn and sink, taking the young shoots to the bottom, where they avoid freezing. In the spring the young plants are released by the decay of the old ones and float to the surface. Some rooted plants have food-packed rhizomes like land plants.

Fast flowing rivers and streams contain only those plants that remain firmly rooted. As the flow slows down a wider variety of plants can establish themselves, in the channel, in the shallows and on the edges. As the water becomes

Leaves and flowers of arrowhead, Sagittaria sagittifolia *rising above Canadian pondweed,* Elodea canadensis. *The thread-like stems and flowers of the latter plant are rarely produced in this country. This photograph was taken on the Somerset Levels*

shallower, through siltation or evaporation, in disused canals or dykes the edge plants encroach further into the centre, eliminating much of the variety. Although plants will vary in response to the rocks, water type and other factors, it is the way in which a plant is growing and its position in relation to the water that are most noticeable. Differences in the acidity of water may mean that alternate-flowered water-milfoil *Myriophyllum alterniflorum* is present rather than the spiked, *M. spicatum,* which prefers richer more basic water but they are both submerged plants living in much the same way.

The main groups of plants are:-

(a) Submerged
(b) Floating leaves but rooted
(c) Free-floating
(d) Emergent but with floating or submerged leaves
(e) Reedswamp
(f) Marsh

In other books (d) and (e) are often looked at together and there is considerable overlap between the last three groups depending on the level of the water, the current and the amount of silt deposited. The sections that follow are not intended as a guide to plant identification: books for this purpose are suggested later and only a selection of species are dealt with below.

Submerged Plants

These are rooted in mud or attached to rocks in up to 3m (10ft) of water, depending on the turbidity. Although some spend all their lives under water, for example autumnal starwort, *Callitriche hermaphroditica,* others like Canadian pondweed flower above the surface. Not all are flowering plants, lower plants such as algae (the same group as seaweeds) also occur. They can form bright green tufts on stones, or grow on other plants and are an important food source for invertebrates. One group of algae form blanket weed. Stoneworts, *Chara* and *Nitella* species, so called because they become encrusted with calcium carbonate, are a small ancient group, not obviously related to other plants, except some of the algae. They occur in streams and canals and are anchored by colourless threads. They are very branched and may be quite dense in suitable conditions. Willow moss, *Fontinalis antipyretica,* another of the lower plants can grow up to 0.5m (20in) long and is our largest moss. It has dark green, bushy, trailing stems attached to stones and is often found on weirs and lock sides with *Rhychostegium riparoides* and other mosses that survive submersion. Willow moss is an important shelter and food source for cranefly larvae and other invertebrates.

The three native species of water-milfoils, whorled, *Myriophyllum verticillatum,* spiked and alternate-flowered show a gradation of their habitat preference. They are unmistakeable with feathery whorls of leaves and flower spikes which grow above the surface. All are perennial, having rhizomes, and can survive as fragments; as an additional safeguard the whorled water-milfoil produces winter buds. Spiked water-milfoil is the commonest, scattered throughout Britain, especially in calcareous water in moderate flows, whereas the alternate-flowered species is more common in the north and west where fast flowing oligotrophic streams occur. The rarer, whorled water-milfoil occurs mainly in south-east and central England in slow flowing water.

The large pondweed family, Potamogetonaceae, exhibits considerable variety, with submerged and floating species occurring in fast or slow, eutrophic or oligotrophic waters. It includes over twenty *Potamogeton* species and their hybrids all with alternate leaves, and the opposite-leaved pondweed, *Groenlandia densa.* Some of the *Potamogeton* species grow to over 1.8m (6ft) in length. The petal-less pale green flowers usually grow above the surface but once the fruits develop, they sink. *Groenlandia* and fourteen *Potamogeton* species occur only as submerged plants in flowing water. They include the common and widespread, curled pondweed, *P. crispus,* and the rare, south-eastern sharp-leaved, *P. acutifolius.* The distribution of some, for example the shining pondweed, *P. lucens,* is limited to base-rich water. Most of the submerged pondweeds have grass-like leaves but some are broad-leaved, providing a large surface area. *P. perfoliatus,* a common lowland species has, as its name suggests, leaves which clasp the stem and they are broader than many of the floating leaves.

The crowfoots, *Ranunculus* species, relatives of buttercups, are another varied group but they go one stage further, for the same species can appear in different forms depending on the current. They all have distinctive white buttercup flowers which when numerous look like foam on the surface of the water. Like the pondweeds some species have only hair-like submerged leaves, others only the kidney or heart-shaped floating leaves and there are a few that have both. River crowfoot, *R. fluitans,* is found in fast-flowing rivers, where it can tolerate heavy spates, but also occurs in lowland streams. It has long tress-like leaves up to 0.3m (1ft) long. Whereas in the south-eastern *R. circinatus* the stiff, fine leaves form a fan shape less than 2.5cm (1in) in diameter. There are three varieties of *R. penicillatus;* one is adapted to fast flowing, highly alkaline waters, one to less rich water and a rare one occurs in canals and slow streams only.

Hornworts, *Ceratophyllum* species, although they have no roots are anchored by modified leaves, at least in the early stages. They are easily detached in a current and float below the surface. Hornwort, *C. demersum,* is more common than the softer *C. submersum* found mainly in south-east Wales, Somerset and occasionally in the east. They are bushy plants with stiff spiky leaves and as often happens the main identification feature, the presence or absence of spines on the fruit, is not much help because fruits are not commonly produced. Canal reservoirs provide still water, which is ideal for some plants that can not survive in flowing water. Shoreweed, *Littorella uniflora*, usually grows in shallow water at the margins but it can survive a depth of 3m (10ft). It is not particularly attractive, being a relative of the plantains, and its inconspicuous flowers are usually only produced when it is out of the water. Obviously canal reservoirs, like the one at Combs in Derbyshire are ideal for flowering as the level of water drops in summer.

White water-lily, **Nymphaea alba,** *in an ox-bow of the River Dove on the Staffordshire-Derbyshire border*

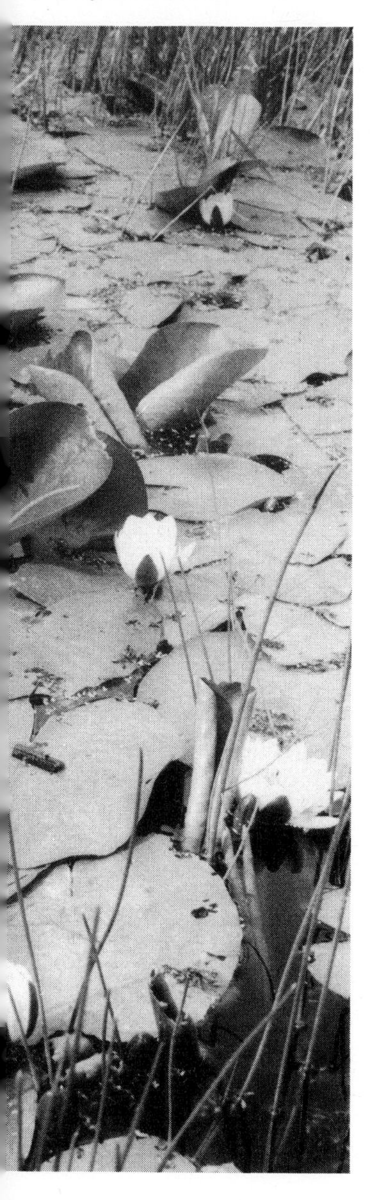

Floating Plants

Some plants with floating leaves are rooted and can survive in faster flows than the duckweeds and other free floating plants. The long trailing stems are vulnerable to bruising and cutting by propellors. They may also have submerged leaves which may be strap-like in contrast to the flat, often large floating leaves and the flowers, like those of the free-floating plants often project above the surface to help pollination. Some groups, like the water-lilies, have their long leafstalk attached in the middle of the leaf to reduce the likelihood of overturning. There are two related species found in flowing water, the more common yellow water-lily, *Nuphar lutea,* and the white one, *Nymphaea alba,* which is found far into Scotland. The unrelated fringed water-lily, *Nymphoides peltata,* occurs in the Fens. Only the yellow water-lily has cabbage-like submerged leaves and in faster water, its leaves become smaller. Water-lilies can grow in water up to 3m (10ft) deep and the leaf stalks can be very long.

The rest of the crowfoots occur in this group, some like Lenormand's water-crowfoot, *R. omiophyllus,* and ivy-leaved crowfoot, *R. hederaceus* have only lobed, dark green floating leaves. Ivy-leaved crowfoot can be found in a wide range of flows and water types but Lenormand's water-crowfoot is a western species extending north to Argyll in oligotrophic, slow streams. The remainder have finely dissected submerged leaves and larger floating ones; the common *R. aquatilis* which occurs in fast water has three types of leaves, dissected submerged ones, kidney-shaped floating leaves and some that are intermediate between the two.

The broadleaved pondweed, *P. natans,* is a common sight in river slacks and canals; although it will grow in deep water it often forms a large part of the floating vegetation in disused canals. It has narrow submerged leaves and can cope with spates, but not a constant fast flow. Its less common relatives include red pondweed, *P. alpinus,* which also occurs in canals and can survive spates, the fen pondweed, *P. coloratus,* and *P. gramineus,* which tolerates some pollution.

The water starworts are a difficult group to identify because their leaves vary so much, depending on the depth and flow of water and the only certain identification feature is the fruit, which is unfortunately not always produced. They do form deep luxuriant green mats in 0.3-0.6 metres (1-2ft) of water and are a source of food and shelter for a variety of animals.

The double life of amphibious bistort, *Polygonum amphibium* sets it apart from other plants which have found ways of living in water, for it can be either an aquatic or a terrestrial plant. The floating mats of hairless leathery leaves which support the attractive dense spikes of pink flowers are attached by long stems to an underwater rhizome. The leaves are hairy when it grows on dry land. There is another plant, the grass fiorin, *Agrostis stolonifera,* which seems at home both in water and on land but the only British grasses which live only in water are the flote-grasses, *Glyceria fluitans* and *G. plicata.* Other members of this large family occur as emergent, reedbed and marsh species.

Fringed water lily,* Nymphoides pettata, *a less common fen species, unrelated to the common yellow water lily

Free Floating Plants

The next group are not rooted, but float at or just below the surface; they can only occur where the flow does not wash them away and they tend to accumulate against obstructions like lock gates and in backwaters. Canals are ideal for these plants and have helped in their dispersal. Sometimes they grow over submerged plants unless their density excludes light, but they also occur in deep pools where submerged plants cannot grow.

Unless pollution has caused a bloom the least obvious components of this group are algae, in suspension. The plankton consists of free-floating species as well as those, such as the green *Chlamydomonas,* which can swim. They are important food for animals such as water fleas, *Daphnia,* which sometimes look green because of this. Young perch feed on diatoms which have hard silica shells, sculptured in various distinctive patterns. There is usually a natural flush of growth in spring and early summer although ones induced by pollution may occur later and block ditches and unused canals, causing de-oxygenation and other problems.

Two species of liverworts, which are in the same class of lower plants as mosses, occur in water, whereas most are found in damp situations. Liverworts usually have three rows of simple leaves, although in some the plant is flat and lobed. However, *Riccia fluitans* has pale green alga-like forked fronds which float and look translucent because of air chambers.

Although it is an introduced species, the water fern, *Azolla filiculoides,* is worth mentioning because it is the only free-floating fern in this country. It comes from the warmer parts of America, so it prefers the southern counties although it can survive in the Midlands in mild winters. It has caused problems in ditches and canals, for example, the Grand Western Canal in Devon, where it

formed a dense carpet across the canal for over a mile. preventing oxygen and light from reaching the water.

Duckweeds are more familiar as sheets of green covering ponds, canals and backwaters, but they are worth a closer look. The plants increase in complexity from the single oval fronds of *Lemna gibba* and *Wolffia arrhiza* via the attached clusters of round fronds of *L. polyrrhiza* and the oval ones of common duckweed, *L. minor* to the partially submerged, branched, ivy-leaved duckweed, *L. trisulca.* All the *Lemna* species have roots, which hang singly from each frond except in the case of greater duckweed, which has several as the name, *polyrrhiza,* suggests. *Wolffia arrhiza* has no roots. Most of them flower infrequently unless they are drying out and only in common duckweed are flowers usually seen. Greater duckweed and *L. gibba* are southern species; common duckweed is naturally the most widespread, reaching Scotland, where it is found with ivy-leaved duckweed in the Forth and Clyde Canal. The Somerset Levels are now an important stronghold for *Wolffia arrhiza* which has disappeared from several sites in south-eastern England in the last 50 years. Two attractive rarer plants, which are related to Canadian pondweed come to the surface in summer and then sink or produce winter buds in order to survive the cold weather. Frog-bit, *Hydrocharis morsus-ranae,* remains at the surface for much of the summer producing new shoots from stolons. The rosette of lanceolate, spiny-edged leaves of water soldier, *Stratiodes aloides,* surfaces to produce similar flowers. Water soldier is rarer, its stronghold being the Norfolk Broads with odd sites in east and central England. Frog-bit is found in similar places, but also on the Somerset Levels and places as far apart as Tyne and Wear and in canals in Shropshire, where there is no pollution.

Another interesting group are the bladderworts which have the unusual habit of trapping small animals. Several plants supplement their diet in this way. Although none of the species is common, the greater bladderwort, *Utricularia vulgaris,* is found more often than the others and has two forms, one living in basic and the other in acid pools and ditches. It has feathery leaves with numerous bladders where the prey is trapped and digested. In autumn the bladders fill with water and the plant sinks. *U. intermedia* and *U. minor* are rarer and have some stems with bladders which are sunk in the mud. They can be overlooked, but are unmistakable when the attractive spikes of yellow flowers appear. Those of the greater bladderwort are bright yellow and the others have paler flowers.

Emergent Plants

These are separated from the previous groups because

Water plantain, Alisma plantago-aquatica *in a typical canal-margin location*

The flowering rush, Butomus umbellatus

their stems and leaves, as well as the flowers, grow above the water, although most of them also have floating or submerged leaves. Emergent plants are the next stage in the transition from truly aquatic plants to the marsh and bank communities; even so they may often be found in deep water amongst waterlilies and duckweeds. They do not form reedswamp or dense mats which build up in the water, although clumps of bulrush, *Schoenoplectus lacustris* or watercress, *Nasturtium officinale* can become quite thick.

Three attractive plants come into this group, the commonest being water-plantain, *Alisma plantago-aquatica,* which reaches further north and west than arrowhead, *Sagittaria sagittifolia* and the unrelated flowering rush, *Butomus umbellatus.* The broad leaves of water plantain gave it its name and are a common sight on canals and rivers in shallow water or on muddy margins. It has some rare relatives, like the ribbon-leaved *A. gramineum* although the narrow-leaved *A. lanceolatum,* which occurs in southern England, is more numerous than the common one in parts of the polluted Sheffield Canal.

Arrowhead, with its unmistakable emergent leaves, forms patches in deeper water than water-plantain; although it does not like very polluted waters it does grow in industrial stretches, such as the Ashton Canal in Manchester. It is not found in the south-west and extreme north of England, most of Wales and only occurs in Scotland as an introduction in the Forth and Clyde Canal. Unlike water-plantain it has unisexual flowers, the males being above the females, and it produces yellow spotted bright blue winter buds on long runners in the mud. The flowering rush is not a rush at all, although the thick, juicy leafless flower stem which can be 1.3m (4ft) high may have caused the confusion. It is not as common as the other two, but it may be overlooked when the bright pink flowers are not present as it often grows singly or in small groups whereas the other species form larger colonies.

The bulrush, with its smooth tall leafless stems may have been one of the plants for which flowering rush was mistaken. However it can reach 2.6-3m (8-9ft) in height. Its flowers, borne at the tip of the stem, are small and reddish brown. This is not the plant that many people call bulrush; this develops velvet-brown cyclindrical seedheads, and is more properly called reedmace, *Typha latifolia.* It was depicted, wrongly, in an illustration of Moses in the bulrushes and the mistake has stuck.

Mare's-tail, *Hippuris vulgaris,* also emerges as a single unbranched shoot, although like bulrush it has simple submerged leaves. There are dense whorls of leaves up the 0.6-1m (2-3ft) stems. There can be confusion between mare's-tail and the horsetails, which is not helped by the similarity of the names. Although it appears as single shoots, 0.6-1.6m (2-5ft) tall, the water horsetail, *Equisetum fluviatile* is often almost devoid of branches. Horsetails are primitive plants, the survivors of a large group which included trees and smaller plants found now as fossils in coal-bearing rocks. They develop a spore cone at the tip of the shoot, in contrast to the green flowers which appear in the leaf whorls of mare's.tail. Water horsetail can become dominant in slow-flowing water. The marsh horsetail, *E. palustre* has more branches but only occurs in damp areas. These two horsetails are more common than mare's-tail and do not have submerged leaves. Wood horsetail, *E. sylvaticum,* which has branched branches and is a northerly species, and common horsetail, *E. arvense* both produce the cone on separate stems and prefer drier habitats. Giant horsetail, *E. telmateia,* which favours spring lines can be up to 2m (6ft) tall; the rarer northern *E. hyemale* is found on shady stream banks and its shorter, stouter stems have rough ridges which were once used for cleaning cooking pots. These ridges and the toothed sheaths where the branches occur are important in identifying the various horsetails.

Two plants which have obviously caused confusion in the past and continue to do so are watercress and fool's watercress, *Apium nodiflorum,* with a third, narrow-leaved water-parsnip, *Berula erecta* adding to the problems. The last two are umbellifers, related to cow parsley, not members of the cabbage family like cress. It is the leaves that bear some resemblance and a careful examination is necessary to reveal the differences, particularly as they frequently grow together, preferring limestone or chalk streams. Watercress, cultivated as a salad plant in clean, unshaded streams has smooth edges to its rounded leaves whereas fool's watercress has finely toothed pointed leaflets and the water-parsnip has coarser teeth. Watercress remains green all the year round which is useful for the growers, but it has a northern relative, *N. microphyllum,* which turns bronze in autumn.

The tall, white-flowered dropworts are also umbellifers, a large group that many people discard as too difficult to identify; yet if the plants are examined closely and the habitat considered they become easier to separate. Fine-leaved water-dropwort, *Oenanthe aquatica,* and river water-dropwort, *O. fluviatile,* have emergent and submergent leaves but only grow in relatively shallow water, usually 0.6-1m (2-3ft) deep although the former can be 2.3m (7ft) tall. Tubular water dropwort, *O. fistulosa,* and hemlock water-dropwort, *O. crocata,* are both totally emergent, growing in shallow water and on muddy banks. This poisonous group has an interesting distribution; fine-leaved water-dropwort favours faster flowing chalk and limestone streams in the south and east, river and tubular water-dropworts occur in south-eastern and central England and hemlock water-dropwort is widespread in the west and north-west where the others are not often seen.

The spearworts (relatives of buttercups), forget-me-nots and speedwells are found in water but also commonly in reedswamps and marshes. Lesser spearwort, *Ranunculus flammula* is widespread throughout Great Britain whereas greater spearwort, *R. lingua,* has a scattered distribution and has been introduced into ponds, canals and streams. Both grow in water although lesser spearwort is more likely to be found in the middle of small slow streams than greater spearwort, which is usually found on the edges of oxbows, canals and ditches. Greater spearwort has large, floppy leaves and its bright yellow flowers can be 50mm (2in) across, twice the size of its smaller relative. The most common water forget-me-not is *Myosotis scorpioides* which has deep blue flowers and grows in 0.3m (1ft) of water or

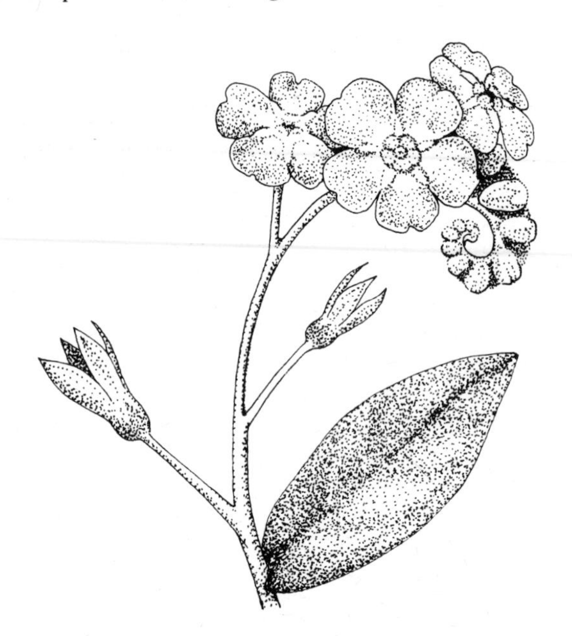

The northern species of water forget-me-not, **Myositis secunda**

on the edges. Like so many plants it has a northern counterpart, *M. secunda* which prefers acid soils.

There are two water speedwells, the blue flowered *Veronica anagallis-aquatica* which is more widespread than the pink *V. catenata.* Marsh speedwell *V. scutellata* has white, pink or pale blue flowers and is more westerly and northerly than the other two. All are more upright than the common fleshy-leaved brooklime, *V. beccabunga* which has eye-catching dark blue flowers and may choke small ditches and streams with its sprawling stems. Great water dock, *Rumex hydrolapathum* does not form swamps but unlike most of the others in this section it has no floating or submerged leaves. The dense clumps of leaves which can be 1m (3ft) tall, over-topped by the typical dock flower-head are common along canals and ditches in the water or on the marshy margins.

Reedswamp

This zone extends into shallow water and there is often a band of aquatic plants on the stream side. All the plants in this group are emergents with vertical stems and leaves. Bur-reed, *Sparganium erectum,* has air-storage spaces (aerenchyma) which help when the water rises and occasionally it has floating or submerged leaves, but it

A typical canal side, with water dock,* Rumex hydrolapathum, *in the water and meadowsweet,* Filipendula ulmaria *among other marsh plants higher on the bank

does create swamp conditions and is more properly placed in this group. The main characteristic of the swamp-forming species, reed, *Phragmites australis,* the reedmaces, *Typha latifolia* and *T. angustifolia,* bur-reed and reed sweet-grass, *Glyceria maxima,* is their growth habit. Their underground stems (rhizomes) and roots form interwoven mats which can withstand some wave action and floods. As a result they come to grow in dense stands which act as a filter for silt carried by the water. The leaf debris and silt accumulates gradually, becoming drier as the swamp builds up.

As all the main constituents of reedswamp are widespread, it is flow, water quality and depth that are important in determining which species occur and with such precise ecological requirements they often grow in almost pure stands. Reed can grow in water 2m (6ft) deep if there is no flow, but in moving water it is rarely in water deeper than 1-1.3m (3-4ft). The stems which bear the purple flower heads may grow to 3.6-4.2m (12-14ft). Pollution or wash from boats reduce it to a narrow fringe in canals and some rivers where conditions might otherwise be ideal. Large reed-beds occur on the Norfolk Broads

.ere they have been traditionally managed to produce .hatching material. The frequency of harvesting influences the quantity and quality of the reed and annual or biennial cuts are best. Studies of reedbeds have indicated that cutting has no adverse effect on their bird-life other than removing potential nesting sites, because new reed growth appears too late to be used for nesting in that year. Provided the harvest leaves enough uncut reed for nest sites, reed warbler, sedge warbler, grasshopper warbler and reed bunting will still be found there.

In disused canals, where it is usually the commonest species, and in polluted water the other grass, reed sweet-grass, which has green flowers, often takes over from reed. It can become dominant and because it has few basal leaves it provides a poor habitat for animals. It grows in shallower water than reed and is more common in the south. It is rare in Scotland and botanists finding it in the Forth and Clyde and Scottish Union Canals were delighted, unlike canal nature reserve managers and restorers south of the border who find it an aggressive opponent. It can survive for several years in heaps dumped on the banks.

There are two species of reedmace: the greater, *T. latifolia,* and the less common *T. angustifolia,* but only the former will replace reed in polluted water. Both produce the brown seed heads which crown many dried flower arrangements. Although the lesser reedmace is, as the name suggests a slighter plant, it is the flower spike that provides a good identification point. The top (male) and lower (female) flowers are separated by a piece of stalk in lesser reedmace but are contiguous in greater reedmace. Both reedmace, and to a lesser extent reed sweet-grass have the disconcerting habit of forming floating mats of rhizomes over quite deep water, so beware!

Branched bur-reed, *Sparganium erectum,* is often found in patches on muddy ground on the landward side of reedswamps but it can also become the dominant plant in the water along the margins of rivers, ditches and canals, both used and disused. It is considered to be the most widely distributed stream plant, the flowers are wind-pollinated with the males above the females, looking like round stars. Unbranched bur-reed, *S. emersum* often floats, rarely forming dense stands and can be uncommon locally. The burrs which contain the seeds are eaten by waterfowl.

Two unrelated species with similar leaves, yellow flag, *Iris pseudacorus* and sweet flag, *Acorus calamus* can also form dense stands in shallow water. The attractive yellow flag is well known, but the introduced sweet flag is often overlooked or mistaken either for its namesake or for bur-reed. Closer examination reveals that the crinkled leaves of sweet flag have an off-centre midrib, unlike the others; the sweet smell from the crushed leaves, which some people liken to tangerines is also distinctive. It rarely produces the

***One of the most beautiful of marsh plants, the yellow flag,* Iris pseudacorus**

The upper, smaller flowers of branched bur-reed, Sparganuim erectum *are male; the lower, female ones develop into the burrs*

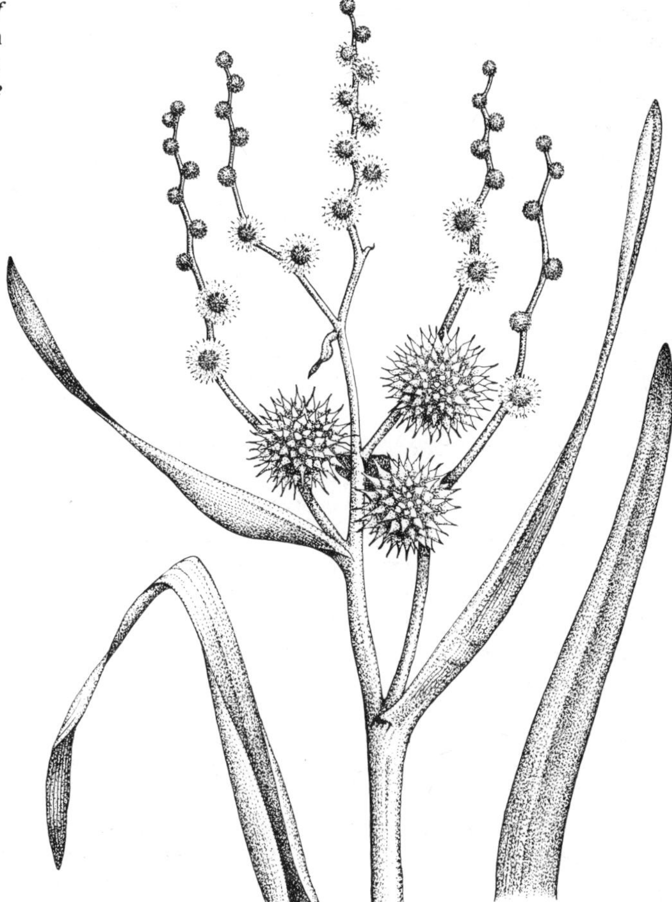

peculiar green flowering spike, but some years, for example, 1980, seem to provide the right conditions.

The greater pond-sedge, *Carex riparia,* and the lesser pond-sedge, *C. acutiformis,* can also form dense patches, either singly or together, building out into slow flowing water. Sedges are a group of plants that are not easy to identify because the flowers are tiny and the fruits are necessary for final confirmation of some species. They resemble grasses, woodrushes and true rushes but they have triangular stems which are never hollow. It helps to remember that sedges have edges! Size is not always useful for distinguishing the greater and lesser pond-sedges, but the latter is more slender and greyish and the lower, female flowers remain erect whereas those of the larger species droop. Several other sedges grow in water but they do not form swamps and are mentioned in the section on marshes and banks.

5 *Aquatic Invertebrates*

The surface of the water casts a screen across a world at least as active and interesting as the terrestrial habitats alongside. Anglers have a limited insight into this world, and perhaps narrowboat enthusiasts are aware of life beneath their feet, but its diversity is largely hidden from the casual observer who can see little through such a turbid and distorted barrier. This chapter describes the most visible and important aquatic invertebrates of lowland waterways, but is by no means a complete survey and is only intended as a summary of the major groups.

Creatures adapted to a totally aquatic life need to have solved many physiological problems, and because of this they often bear little resemblance to land-based relatives. Some invertebrates evolved in marine environments and then made the difficult transition to freshwater conditions. Many evolved on land, however, and took to the water to make use of feeding niches that were unexploited. Different families of invertebrates developed in different ways, their species radiating out to specialise in the conditions produced by varying types of water habitat. Some of the complex influences on their distribution have been considered in chapter 3. It is possible to mark the zones of a river by the invertebrates present, but this is not altogether successful because of the localised nature of their colonies. A more realistic method of biological zonation is by the identification of the fish within particular zones. This is described in the following chapter, where four idealized zones are recognised. For invertebrate animals, the upper reaches of rivers are characterised by a limited range of species that are able to cope with the rapid current and the sparse food supply. The middle and lower reaches contrast in several fundamental ways. The water is a little warmer and its slower movement allows aquatic plants to root and grow. Thus invertebrates do not have such an abundant supply of dissolved oxygen, but are able to detach themselves from pebbles without being swept away, and are able to feed on an abundance of decaying vegetable matter brought down and deposited by the decreasing current. Also, new habitats are provided by the

▷ ***An adult great diving beetle,* Dytiscus marginalis**

Cyclops, one of the most abundant fresh-water crustacea; the large oval egg sacs at the side of the body can be seen clearly

A curiously-shaped crustacean, the fish louse,* Argulus foliaceus, *which inhabits canals and slow-flowing rivers where it is parasitic on fish

King's Sedgemoor Drain, on the Somerset Levels, is a vital drainage channel and is regularly cleared of vegetation. Even so, it is prolific in aquatic life

silt and mud, and by an increasing range of aquatic plants.

The best conditions are provided by slow-moving waters, rich in nutrients and with little regular interference. Such habitats are rare, and are probably encountered most often in disused canals and fenland drains where spraying has not been encouraged. Most lowland waterways are compromised by one or other deficiency, but even so a wide range of invertebrates occurs on almost every river or canal system. Several groups, like protozoans and rotifers have been excluded from the following summary because they cannot be seen without a microscope, but it should be remembered that they too are a vital link in the aquatic food chain.

◁

Almost ready for flight*: Aeshna juncea, *the common hawker dragonfly, drying its wings. The old nymphal case will stay hooked to the bur-reed stem until rain or wind dislodge it

Sponges: Porifera

It is difficult to believe that sponges are animals; they do not move and can be confused with blobs of algae or fungi. In fact they were once placed in their own kingdom, neither plants nor animals, but are now recognised as very primitive animals resembling colonies of protozoa. Eight species are recognised as British, but only two are at all common. *Spongilla lacustris,* the pond sponge, prefers rather deep water but is plentiful above canal lock gates, where it thrives on submerged structures. It is a pale off-yellow or green colour and has fingered projections that give it a rather sinister appearance. The other species is *Ephydatia fluviatilis,* the river sponge, which avoids rivers almost completely and usually grows on the undersides of stones like a thin, undercooked pancake.

For all their lack of charm, sponges are interesting creatures, requiring lime-free water with a supply of silicates from which they construct their skeletal structures. They were once used as a dried powder to cure rheumatism, but today they are almost totally ignored and seem to play a very minor role in the ecology of freshwater. They feed by filtering tiny food-fragments from water that is pumped through their pores. Sponges only grow during the summer months, and are invaded by several invertebrates including the larvae of lacewings belonging to the family Sisyridae. These feed on the body-fluid of the host sponges and then, when full-grown, swim back to the bank to pupate.

Hydra: Ceolenterates

Most children are forced to draw hydra at some stage of their school career, yet few ever see the living creature. The reason for this is quite simple, for although at least three species are quite common and of a reasonable size (ie about 2cm with even longer tentacles) they are capable of contracting to tiny nodules. They do this when disturbed, and so under normal circumstances pond-netting produces nothing resembling the diagrams in school text-books. The answer is to collect pond-weed and let it stand

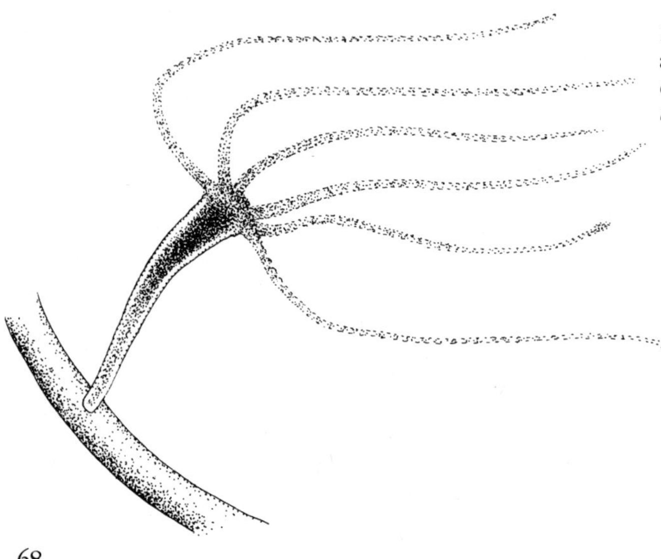

The brown hydra,* Hydra oligactis, *is up to 2cm long but can contract to only a fraction of this when disturbed

in a jar for a day or two. Any hydra will then overcome their trauma and open out to reveal themselves as emaciated but insatiable predators of small fish-fry and water fleas.

The commonest species is probably *Hydra oligactis,* the brown hydra, which has tentacles several centimetres long and is particularly slim and elegant. *Chlorohydra viridissima,* the green hydra, is distinctive because of its association with a minute alga that gives it its beautiful emerald colour. Other species occur, but there is some uncertainty as to exactly how many there are. Their marine relatives, the jellyfish, sometimes find their way into estuaries, but the only other Ceolenterates seen in freshwater are the medusae released from time to time into industrially-heated waterways.

Flatworms: Platyhelminthes

Flatworms are extremely common and to some extent the different species can be used to zone a river system. Unfortunately identification of the eleven British species is extremely difficult and this makes such an exercise more of a problem in the use of scientific keys than in the isolation of river reaches. Flatworms are rather ugly creatures, extremely compressed and squared off at the front end, but otherwise lacking any distinctive features. They vary in size, but the species most likely to be found on canals and slow rivers are *Dendrocoelum lacteum,* a white creature up to 25mm long, and the much smaller *Polycelis nigra,* which is black and can be found abundantly among submerged vegetation.

The most notable attribute of flatworms is their ability to regenerate themselves from small pieces. If cut into four or five sections, each section will develop its own central nervous system and become a new flatworm. They are extremely primitive animals, but have a rudimentary brain and are important predators, attacking crustaceans and other water invertebrates. They also act as scavengers on any dead fish or mammals, coming out at night to hunt, using their sense of smell rather than sight to locate their prey.

Water Snails: Gastropod Molluscs

Gastropods, quite literally, have their stomach where their foot is, and scrape up food as they glide along. Most of the thirty-six freshwater species resemble land snails, and are separated into two groups, those with gills and those with lungs. The Prosobranch group breathe through a kind of gill and usually prefer well-oxygenated water. They can be easily recognised by their 'operculum' — a hard plate used to close the entrance to the shell and protect the occupant. The other group, breathing mainly through a lung cavity, are called Pulmonates. They probably originated from land snails and can tolerate nearly stagnant water because they still rely on oxygen from the surface. Lowland rivers and canals contain representatives of both groups, but the Prosobranch (or 'operculate') species are not found on disused canals where there is no water flow.

The most beautiful of the water snails is probably *Theodoxus fluviatilis,* the Nerite, an operculate species less than a centimetre across but with a wide, yellow-brown shell and a series of pale purple, oval markings. It is found wherever there is a little lime in the water, and is particularly common on lowland rivers like the Avon which have a good zone of emergent vegetation. Even at its most numerous, however, it is never as abundant as *Potamopyrgus jenkinsi,* Jenkins' spire shell, a tiny dark brown creature which used to be associated with brackish (ie slightly salty) water, but has now invaded most river systems and is one of the most dominant species. It belongs to the family Hydrobiidae, which also includes several other small brown species. *Bythinella scholtzi,* which grows to a maximum of about 3mm, belongs to the same undramatic family, but is of particular interest because it is only found in two sites in Britain — one in a dock in Stirling and the other in canals near Manchester.

Of all the operculate snails, the river snails or freshwater winkles are probably the best known. There are two British species, of which *Viviparus viviparus* is the most widespread, being found on most rivers and canals (for example the Trent and Mersey near Derby). Its presence in waterway habitats is easily detected by looking for freshly-dredged sections, where the old shells are sturdy and lie on the surface for some time before breaking down. The shells are large (over 3cm) with a series of three broad dark stripes. Among the Pulmonate snails, the best-known is

The river snail,* Viviparus viviparus, *a sturdy and well-marked species with a strong operculum to close the entrance to its shell (up to 40mm)

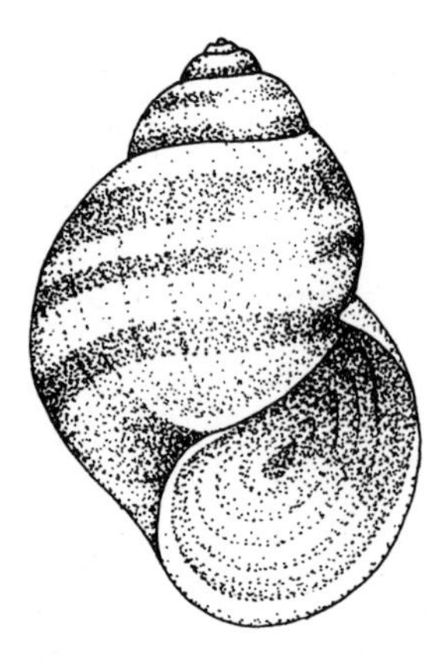

The largest ramshorn snail,* Planorbarius corneus, *is characteristic of lime-rich rivers and lakes. (30mm)

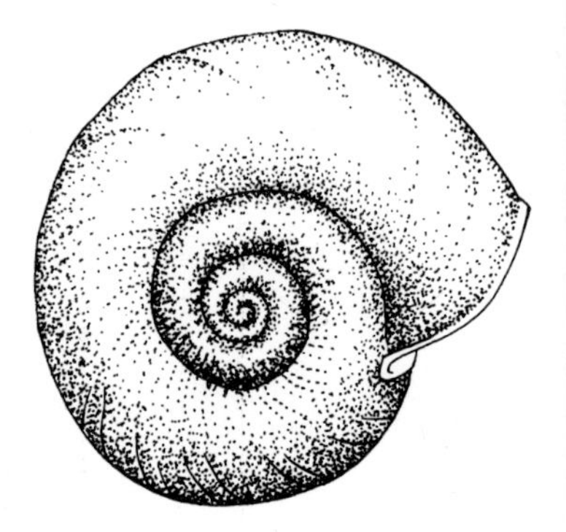

certainly *Lymnaea stagnalis,* the great pond snail, which can grow up to 5cm and has a distinctive tapering spiral. It is one of the few snails to show carnivorous tendencies and will attack injured fish as readily as algae. Several other species of pond snails occur, all of smaller size but of a pointed shape and dark brown in colour. *Lymnaea peregra,* the wandering snail, is the commonest and is found in almost every kind of habitat.

Ramshorn snails also belong to the Pulmonate group and differ in structure from the others, having a series of flattened whorls that makes them look like little ammonites. They are common in running water, the largest being *Planorbarius corneus,* the great ramshorn, which prefers hard water and grows up to 3cm in diameter. Of the remaining thirteen species, *Planorbis acronicus* is of special interest because it is restricted to the tributaries of the Thames. It is difficult to judge whether such species have contracted in range and were once widespread or have always been confined to very specific habitats. Freshwater systems are particularly susceptible to isolation; conditions downstream may deteriorate and leave invertebrate populations incapable of expansion.

Mussels and Cockles: Bivalve Molluscs

Twenty species of bivalves occur in freshwater, sixteen of which are small 'pea-shell cockles' belonging to the genus *Pisidium.* Another genus, *Sphaerium,* contains four 'orb-shell cockles', while the remaining species include the introduced *Dreissena polymorpha,* the zebra mussel, and the true freshwater mussels. Of these *Margaritifera margaritifera,* occurs in fast-flowing water and was once used in pearl fisheries in the north and west, while the other four are all found in slow-moving water. The biggest species is *Anodonta cygnaea,* the swan mussel, which measures up to 14cm (5½in) across and is common along lowland rivers and canals. It is easily confused with *A. anatina,* the duck mussel, which, as the name suggests, ought to be smaller, but in fact the two are very similar and can be separated only with difficulty.

Bivalves live in the muddy substrate with the two halves of their shell slightly apart and two syphons protruding into the clean water. One syphon sucks water in and the other blows it out, having first sieved out food in the form of algae or micro-invertebrates. The early lives of most freshwater bivalves are spent attached to the fins and tails of fish, where they live as parasites encased in small cysts. This immature stage lasts for about three months after which the adult mussels detach themselves and descend to the silt and mud. There they are an important food source for mammals and birds, particularly for species like the tufted duck which must consume vast amounts of pea-shell cockles during the winter months.

Worms and Leeches: Annelids

A great many worms inhabit freshwater. Some, like roundworms (nematodes), are extremely small and have not been included in these descriptions, but the Annelids are all quite large and are particularly common in slow water systems. They are separated into two classes, the Oligochaetes and the Hirudines, the former including earthworms and potworms, and the latter the leeches. Many of the Oligochaetes resemble ordinary earthworms and live in much the same way, in the mud along the waters edge. The largest species belong to the family Lumbricidae, and grow to about 8cm (3in) in length, whilst the potworms include about one hundred and thirty species and are never much more than a centimetre long.

Tubifex worms are especially noteworthy in that they provide a most important food-source for fish. They are able to live in oxygen-deficient water and are usually red in colour because their bodies contain haemoglobin. This enables them to 'fix' oxygen from comparatively stagnant water, but in order to absorb as much as possible, the worms are obliged to protrude their tails from the tubes in which they live and wave themselves around. Areas of mud under shallow water can sometimes disappear under a pink haze as thousands of these worms compete for both food and oxygen.

Only fourteen species of leech occur in Britain, and while the majority are blood-sucking parasites, a few are predators of other invertebrates. The largest species to be found along rivers and canals is *Haemopsis sanguisuga,* the horse leech, which can extend itself to about 15cm (6in), but is incapable of attacking anything bigger than a young frog. Most other species are small and are associated with fish or (mainly) invertebrates like water snails. They are most common in still or slow-flowing waters where they are able to swim freely or hang from aquatic vegetation while waiting for hosts. Their strong suckers would certainly prevent them from being washed away in the upper reaches of rivers, but most species are incapable of attaching their eggs to anything that could protect them sufficiently in strong currents.

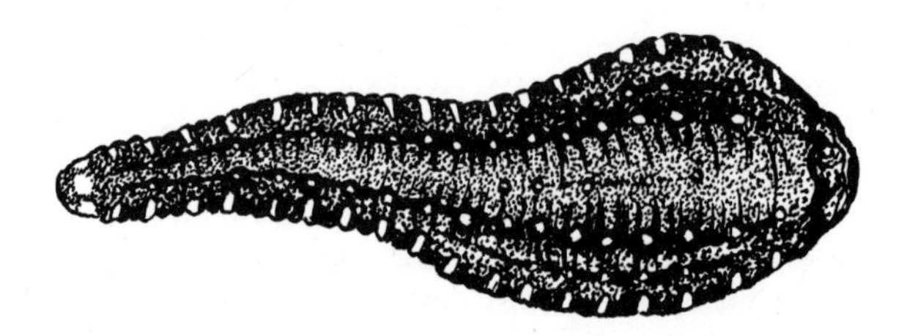

Leeches are particularly numerous along slow-moving rivers and canals. This species,* Glossiphonia complanata, *preys on molluscs and varies from pale green to dark brown. (15+mm)

The most unpleasant life-cycle is provided by the species *Theromyzon tessulatum,* which attaches itself to the nasal cavities of waterfowl — sometimes in sufficient numbers to cause the death of young or unhealthy birds. The only species known to suck the blood of humans in this country is *Hirudo medicinalis,* the medicinal leech which is now very rare because of its popularity in the early nineteenth century. Few naturalists would make out a case for its widespread reintroduction, however, as it is considerably less attractive than other persecuted animals like the osprey or large copper butterfly. A new species, *Haementeria costata,* was discovered in Britain in 1979 and is known to suck human blood in other parts of Europe, but so far it has not been recorded as doing so in this country.

Crustacea

Among the free-swimming and benthic (ie bottom-dwelling) invertebrates, the crustacea are among the most important, both in total numbers as bio-degraders and consumers of plant life, and as food for other animals. A jam-jar full of canal water will almost certainly hold dozens of tiny semi-transparent creatures, including Cladocera, Copepods and Ostracods. Unfortunately while some of these, like *Daphnia pulex,* the common water flea, may measure 3 or 4mm across, the vast majority of our 300 species are so small that they require at least a low-powered microscope to see very much of them. *Cyclops,* belonging to the Copepod subclass, is just about visible as a pear-shaped creature with two egg-sacs attached to its sides. It occurs in vast numbers and must, like the water-flea, be one of the most important fish-foods available in fresh water. The Branchiura is a much easier sub-class to describe since it contains just two species, both nearly a centimetre in size and belonging to the genus *Argulus.* These are the fish-lice; curious dish-shaped creatures which swim around until they come into contact with a fish. They then attach themselves and suck the blood of their host. The common species of waterways is *A. foliaceus* which seems to be particularly numerous in canal basins.

The most easily-observed of the crustacea belong to the sub-class Malacostraca, which includes the freshwater shrimps and the freshwater lice. Neither name is appropriate, and the latter are sometimes called water-slaters to clarify the fact that they are in no way parasitic. Freshwater shrimps, called Amphipods, are easily recognised by their laterally-compressed bodies. When caught they seem to spend all of their time swimming round and round on their sides, males and females often in tandem, and die very quickly in captivity since they require clean running water. The majority of the British species are

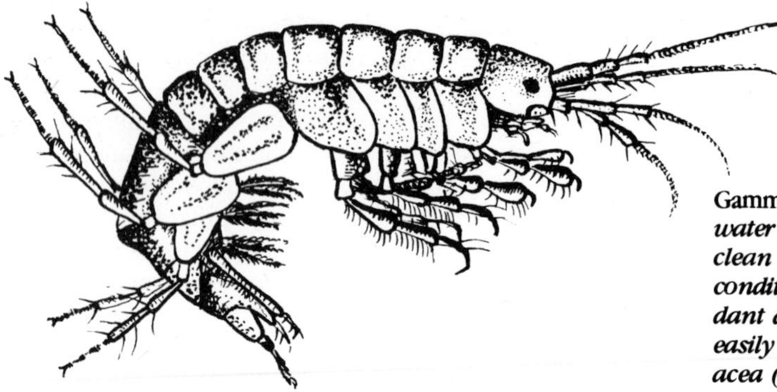

Gammarus pulex, ***the common water shrimp, is an indicator of clean flowing water. In such conditions it can become abundant and is certainly the most easily recognised of the crustacea (15-20mm)***

associated with brackish or salty water, and the only abundant freshwater species is *Gammarus pulex,* which acts as an important scavenger in lowland rivers. *G. tigrinus,* which, as its name suggests, is striped and therefore easily recognised, used to be considered an estuarine species but has penetrated inland to canal systems in the Midlands. Perhaps such canals were once made salty by industrial processes, but today the striped water-shrimp is still spreading and is particularly abundant in the overflow channels of canal locks.

Corophium curvispinum is an even more extreme example of the expansion of an estuarine crustacean into canal systems. It resembles a water shrimp in general appearance, but inhabits mud burrows rather than open water-weed. Most books still describe *Corophium* as characteristic of brackish water, but it has now invaded lock walls and channels on most of the major canal systems, including the Trent and Mersey, Grand Union, Oxford and Worcester and Birmingham. Doubtless it has become an important food-source for fish able to suck the creatures out of their protective burrows, but its abundance in shallow overflow channels suggests that it has few natural enemies capable of capitalising on the rapid change in distribution.

The slaters or water-lice have bodies compressed the opposite way to freshwater shrimps and belong to the group called Isopods. They are related to woodlice and resemble them closely in appearance, having a series of flattened dorsal plates and 7 pairs of legs. They feed either on the bottom mud or among aquatic vegetation, scavenging for scraps of food and being preyed on themselves by a wide range of larger animals. The commonest species is *Asellus aquaticus,* which is abundant in canal and slow river water, while *A. meridianus,* seems restricted to rivers. Slaters can tolerate less clean water than can *Gammarus,* but the presence of *Asellus* is by no means an indication of pollution and the two are separated by habitat rather than by water quality.

The biggest of all the crustaceans are the Decapods. This

group includes the crabs and lobsters, but in Britain the only indigenous freshwater species is the crayfish, *Austropotamobius pallipes*. This looks like a miniature lobster and measures up to 10cm (4in) in length: an impressive and interesting nocturnal predator. Surveys have suggested it has become much more restricted in range, but it is still found along most clean river systems and it has even been found along some urban waterways.

Taken together, crustaceans exhibit incredible diversity, not only in size and structure but also in their ecology. They range from simple herbivores to scavengers, parasites and predators, but they have not achieved such success on land and are almost entirely restricted to water habitats.

Insects

Most aquatic invertebrates take in oxygen by diffusion, an inefficient system in certain respects, but one which releases them from the obligation of having to return to the surface every few minutes to breathe. Adult insects take in air through a system of small holes called spiracles, which means that any water-based species require both access to the surface and a means of storing air around the spiracles. As with the pulmonate water-snails, this air-breathing habit is a sure indication of land-based ancestry, and only two insect orders have a significant number of aquatic species present in the adult state. This does not mean that many other insects do not make use of the habitat in their developing stages, however, and many insect larvae can breathe by simple diffusion or through gills. These are considered separately in chapter 10 because their flighted adults are particularly attractive and well-known inhabitants of waterside vegetation.

The most noticeable insect larvae in freshwater belong to the mayflies, which are extremely active and very numerous. Dragonflies and damselflies larvae are larger, but are less active and much more retiring. Unlike mayflies they are predatory and are therefore never as common or as easily caught. The other insect larvae to draw attention are the caddisflies, because they knit a case of plant or stone sections around themselves and crawl around rather like hermit crabs. They are scavengers, taking plant and animal food, and are often locally abundant in both fast and slow running water. Most insect larvae can be recognised as such because they have 6 legs, but fly grubs usually have none, and the immature stages of many other invertebrate groups bear little resemblance to their adult stages and can be easily confused.

The most important insects with aquatic adults are beetles and bugs. Beetles (Coleoptera) are easily recognised because of their rounded shape and the presence of hardening wing-cases. The most famous water-beetle family is the Dytiscidae, which includes many large carnivorous species like *Dytiscus marginalis,* the great diving beetle, a fearless and aggressive animal which will attack anything remotely edible. The majority of predatory beetles are quite small however, and restrict themselves to crustacea or insect larvae. The family Gyrinidae comprises the whirligig beetles: curious insects, living gregariously on the surface of the water and spending endless hours gyrating in small circles. Whirligigs are carnivores with two pairs of eyes, one positioned above the surface to look for any potential danger and the other below to search for prey. Like the Dytiscidae they are well-adapted to water life, having oar-shaped legs essential for fast movement through the water.

One of the most voraceous of aquatic predators is the great diving beetle* Dytiscus marginalis. *The larva (illustrated) is as fearless as the adult and will attack newts and small fish. (40mm)

The biggest and bulkiest of all the British beetles is *Hydrophilus piceus,* the silver water beetle, which can measure almost 5cm (2in) long. The family Hydrophilidae are not carnivourous in the adult state, however, and so this huge species has had no need to develop the speed and agility of the *Dytiscus* species. It is a herbivore, browsing on dense aquatic vegetation and is becoming increasingly rare as suitable habitats diminish. Old canals are ideal, but despite their functioning wings most of the larger beetles seem reluctant to disperse widely and are slow to colonise new waters.

Bugs (Hemiptera) resemble beetles in some respects, but their wing-cases are not completely rigid and they are rarely so oval in shape. Their life cycle is different too, for there is no distinct metamorphosis and the larvae grow into adults by a series of gradual skin changes. Water bugs seem much more mobile than beetles, and are among the first colonists in freshly-dug ponds. This particularly applies to the water boatmen, the Notonectidae, a family including 4 species belonging to the genus *Notonecta.* All are voracious predators and can swim very rapidly by means of their keeled hind legs. Their larger size and habit of swimming upside down distinguish them from the more numerous lesser water boatmen of the family Corixidae. These are scavengers, and canals and rivers can develop complex colonies containing several different Corixids, all with slightly different ecological niches.

Old canals are particularly attractive to water bugs because of the abundance of aquatic vegetation. *Nepa cinerea,* the water scorpion, is very common and is a large, slow creature up to 3cm in length. The name is derived from its possession of pincers at the front end and a long pointed tail at the back, giving it a vague scorpion-like outline. In fact it is completely harmless. A much rarer relative is *Ranatra linearis,* the water stick insect, which is twice the length of the water scorpion but very much thinner. It is a southern species, supposedly restricted to deep, still waters but certainly found along at least one of the Midland canals where there is ample emergent vegetation.

Apart from the bugs that live underwater, there is a group adapted to life on the surface, the species spending their time skating along in search of invertebrate prey. This group includes the Gerridae or pond skaters and the Veliidae or water crickets. Both are common along canals but the pond skaters include a wider range of species and can be identified by their much longer legs. They usually prey on insects falling into the water from overhanging vegetation, and can move very rapidly over the surface-film to investigate potential food. Like most other bugs, they kill their prey by means of a dagger-shaped beak that injects toxin and is then used to suck the victims dry. The only other surface-dwelling bug likely to be found on slow waters is *Hydrometra stagnorum,* the water-measurer, which paces slowly over the water in search of static prey and is so thin that it looks like a short bristle with legs. Like all the 'skating' species it had fringed legs and body segments that keep it clear of water-tension and allow it to move freely over the surface.

The interrelationship between aquatic and terrestrial insects is very complex and the habitats hold many species reliant on a combination of factors from both systems. Even the most aquatic of insects may emerge into the air at some stage, either for a brief period or as an integral part of the life cycle. The world within a river or canal may be a sealed unit to many invertebrates, but to insects at least there is always an escape route.

Spiders and Mites: Arachnids

Only one spider is aquatic, which is surprising considering the success of these invertebrates in most habitats and the numbers found in marsh and riverside sites. The water spider, *Argyroneta aquatica,* is quite a large species measuring about 10mm in length, and is found on slow-moving or still waters. In most respects it is a typical hunting spider and its adaptations to an underwater life are very superficial. It breathes air, but instead of making continual visits to the surface it spins an underwater platform of silk and takes down a supply of air, a little at a time, attached to the coat of hair around its abdomen, Each consignment is released under the platform to make a bell-shaped dome, and enough oxygen is soon available in the dome for the spider to remain submerged indefinitely. In fact the oxygen supply replenishes itself by diffusion from the surrounding water; an extremely effective strategy and the whole process is a considerable advance on most other aquatic creatures that rely on air from the surface. The only problem is that anything more than a minimal flow of water might detach the air-bell.

The water spider is large but dull in colour, yet the other water arachnids, called water-mites (Hydracarina), are quite the opposite. There are over 200 British species, rarely more than 3mm in length but often bright red or cream-coloured. Most are free-swimming though the immature stages of many are spent as parasites on water beetles or bugs. They are difficult to identify and this may be why they are ignored by many books on pond life; most are oval in shape without an obvious head, and their progress through the water is slow considering they are predators and need to catch small crustacea. They reach their greatest abundance where the water flow is limited, and canals once more provide very suitable habitats. In fact, aquatic invertebrates in general are especially numerous in disused canals and, as farm or village ponds decrease, such waterways are more and more important.

6 *Fish*

No other group of wild animals has endured such a pervasive and meddling influence as fish. Man has altered their distribution and ecology so fundamentally that many are no more natural than free-range hens, and any summary of their biology has to take account of this fact. Fifteen to twenty species inhabit most of the major river systems in Britain; the total depends on many factors including those common to all aquatic life such as temperature, current and nutrient quality. But two additional factors are of particular significance. Many of the richest and most extensive river systems drain into the southern North Sea and were once part of the Rhine. By being part of such a complex and interrelated network, such rivers as the Ouse and Nene were accessible to a wide range of fish species. Conditions were favourable and colonisation must have been quite rapid in inter-glacial and post glacial times, so when Britain finally severed itself from Europe, some 7,500 years ago, these eastern rivers had more than their share of fish. Some of these, like the burbot, *Lota lota,* probably found the newly isolated systems too restricted and the advance of man too

The middle reach of a river (the Teme near Little Hereford) is stony but rich in plant and insect life : an ideal habitat for the dace, Leuciscus leuciscus

detrimental; there seems little doubt that this species will follow others into extinction. But while indirect influences like pollution have prevented such species as the sturgeon, *Acipenser sturio,* from re-establishing itself in our lowland rivers, the artificial introduction of sporting fish has more than redressed the balance. In addition, the cutting of canals and drainage channels has allowed species to disperse from eutrophic eastern rivers to impoverished systems draining into the Atlantic.

Many attempts have been made to classify rivers by the species of fish that inhabit them, and to divide different 'reaches' of rivers according to their dominant species. This has proved reasonably successful on the continent, but for the reasons stated above many British systems lack key species or have confusing mixtures due to artificial introductions. Four zones are usually recognised, and perhaps these are of some relevance since they provide an idealised picture, not only for the fish fauna but for other plants and animals. These are:

(a) trout zone
(b) grayling zone
(c) barbel zone
(d) bream zone

This book deals with lowland rivers and canals, and in this context, the barbel and bream zones roughly correspond to the middle and lower reaches that form the lowland sections of waterways. The fast-flowing primary reaches of rivers, cold and nutrient-poor (ie oligotrophic), are certainly characterised by trout in Britain, but introductions have been so widespread that the presence of trout is more an indication of fly-fishermen than of a classic zone. To some extent this is also true of the second zone, for the grayling, *Thymallus thymallus,* is a favourite among anglers; a sporting fish typical of beautiful and clean waters. Its sensitivity to pollution makes its distribution rather patchy however, to the extent that some authors consider the minnow, *Phoxinus phoxinus,* to be a more appropriate symbol for this fast-flowing reach. Even so, the grayling's presence represents the cut-off point below which the middle and lower zones take over with their much richer fauna and flora. The map on page 83 superimposes its distribution onto that of the bream, *Abramis brama,* the classic species of the lower reach. There is some overlap, mainly caused by artificial introductions of the grayling, but the two exhibit a surprisingly clear demarcation. Between these extremes, and not so confined to text-book conditions, live the vast majority of species. They demonstrate a gradual transition from fast, cold streams to slow turbid rivers. The grayling may not survive where the bream is found, but most fish are adaptable and their ranges overlap considerably. They are the staple angling stock for all of Britain's rivers and canals.

The Middle Reach

Three fish typify the maturing zones of rivers, often of moderate current and stony substrate, but with rooted aquatic plants and a supply of invertebrate food. These are the dace, *Leuciscus leuciscus,* the chub, *Leuciscus cephalus,* and the barbel, *Barbus barbus.* A typical habitat for dace is clean water with a good current where schools of fish can feed extensively on flying insects. Although the dace rarely exceeds 25cm (10in) in length it is popular with anglers and is a slim and attractive species. The chub is closely related, but it is twice the size and has a much broader head. This makes it seem a heavily built fish without the elegance of the dace. It is, however, an important species, spreading further up and down the river and is a vital food-source for river predators. It is often noticed by casual observers from bridges, because the young fish live in close schools and often swim near to the surface. The mature chub has a solid meaty look, but is so bony that it is rarely eaten by humans.

Where the river begins to broaden and the substrate is of gravel rather than stones, the barbel makes its appearance. Since it feeds nocturnally and usually stays close to the bottom its presence is hardly noticed from the river bank. It is an extremely important sporting fish however, and fights hard for its freedom. It can grow to nearly a metre in length, and although it has a rather tubular shape (like an overgrown gudgeon, *Gobio gobio*) it is a very attractive

A day's catch on the Worcester-Birmingham Canal near Offerton. Three good sized chub, **Leuciscus cephalus,** *together with a few roach,* **Rutilus rutilus** *and gudgeon,* **Gobio gobio**

green/gold colour. The barbel used to be quite a local species, but has been introduced into new waters and is gaining popularity among angling clubs.

These classic species are all members of the Cyprinidae, the carp family that includes the majority of well-known freshwater fish. Several other families are represented in the middle reach, however, including the bullhead, *Cottus gobio,* and the stoneloach, *Noemacheilus barbatulus,* both foraging species at home wherever there is a good current. Among the carp family itself, many species of the lower reach can be found well upstream, including the roach *Rutilus rutilus,* the bleak *Alburnus alburnus* and the gudgeon. The apparent confusion between middle and lower reach specialists arises because fast-flowing sections of rivers may contain deep hollows and backwaters suitable for slow-water species. Alternatively, fish adapted to a strong current are quite capable of moving downstream, as spawning seasons and feeding conditions dictate. The actual food spectrum may be quite wide, but for the Cyprinids this usually comprises either semi-aquatic insects taken from the surface (in the case of the bleak, dace and the chub) or benthic invertebrates picked up from the substrate (in the case of the barbel and gudgeon).

The fish-predator of this part of the river is the pike *Esox lucius,* which again is a species of wide tolerance occurring wherever there is aquatic vegetation and sufficient prey.

The Lower Reach

As the current eases and the substrate turns to silt and mud, dace and barbel decrease considerably in numbers. Chub may still be plentiful, however, though they are no longer dominant and are usually outnumbered by roach and (on suitable rivers), bleak. The bream, another of the Cyprinids, gives its name to this final section of the river, and is particularly well-adapted to turbid, rich waters. It is a deep-bodied species, up to 60cm (24in) in length and a golden brown in colour. Its principal food consists of Chironomid midges, *Tubifex* worms and other mud-dwelling invertebrates, all sucked up by means of a 'protractile' mouth that unfolds into a kind of funnel. A closely related species, the silver bream, *Blicca bjoerkna* also feeds on invertebrates and occurs in the same habitat zone. It is only half the size, however, and is more local in distribution; although anglers are not fond of it, direct competition between the species is negligible. As with most other organisms that seem to feed on a similar food source, there are subtle differences in strategy that minimise any over-harvesting of a limited resource: the silver bream takes a high percentage of its food from vegetation rather than from the bottom-mud.

Two other members of the carp family should be

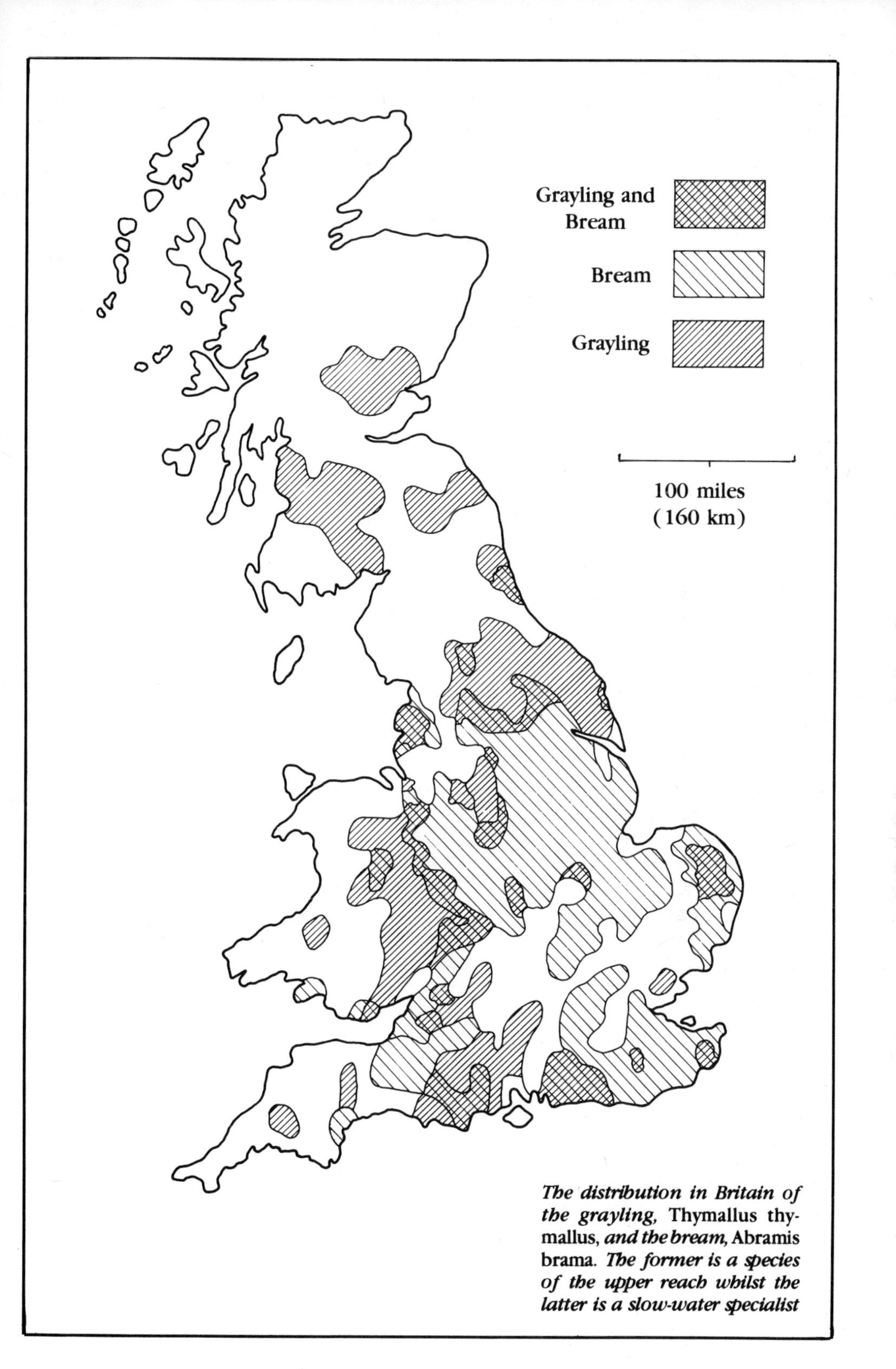

The distribution in Britain of the grayling, Thymallus thymallus, *and the bream,* Abramis brama. *The former is a species of the upper reach whilst the latter is a slow-water specialist*

considered in the context of the bream reach. The first is the tench, *Tinca tinca,* and the second is the carp itself, *Cyprinus carpio.* Both are still-water specialists, particularly the tench which is found in very dense aquatic vegetation and can survive in water very low in oxygen. It is more at home in backwaters and lakes (like the rudd, *Scardinius erythrophthalmus*) but is often introduced into lowland rivers. The presence of the carp is entirely due to the activities of man: it is an alien species but is so popular with some angling groups that it is now a common component of slow, weedy rivers. The introduction of alien species is not always successful, and over-enthusiastic anglers have caused several ecological disasters over the past century. A good example, though on a limited scale, is provided by a member of the Percidae: the perch family that includes the perch *Perca fluviatilis,* the ruffe, *Gymnocephalus cernuus* and the zander, *Stizostedion lucioperca.* While the first two are fully indigenous species, the zander is an east-European import and has proved a troublesome addition to the fauna of the Cambridgeshire Ouse. It is a predator of other fish and has caused a rapid imbalance in what had been a comparatively stable food-web. The perch is also a predator, but by being a native component of slow-flowing waters it has had thousands of years to develop a stable relationship within the biotope, and rarely endangers the balance of prey populations. When young it takes a similar range of food to the ruffe, ie small insects and crustaceans, but as it matures it transfers to sticklebacks, young Cyprinids and other small fish. It is a popular species with naturalists and country-lovers, not because it provides good sport or because it grows to a fair size, usually up to 40cm (16in), but because it is instantly recognisable, with an attractive green/brown upperside and distinctive dark bars and red ventral fins.

The fish that most children hope to see on their riverside walk is a pike. Again it is distinctive, but this time it has the added reputation of being a 'freshwater shark' — totally unjustified but fostered by various folk tales and a recent film about a killer pike. To naturalists it has similar qualities to many of the birds of prey: a solitary hunter, relying on the element of surprise, and always seen when least expected. Apart from the pike, however, the only predators capable of tackling medium-sized fish prey are mammals and birds. In urban or heavily disturbed situations, the lack of these external predators may well result in an unusually high population of healthy pike and perch. In some waters the reverse situation can develop, and a large population of undernourished perch can occur where there is little prey. However this situation does not develop where there are adult pike about, since these will control any surplus stock — not only perch but also young

The best-known of all canal fish is the roach, **Rutilus rutilus,** *but it is also an important inhabitant of the middle and lower reaches of most English rivers*

pike. The ideal natural state is for there to be a large number of herbivorous and omnivorous species, a few perch and one or two pike. This pyramidal structure, so often demonstrated in ecology text-books, rarely works with fish populations because of the influence of anglers.

The lower reach of any river must at some stage make the transition into an estuary, and the introduction of salt water alters the whole ecology. While some fish, like the eel, *Anguilla anguilla,* and the flounder, *Platichthys flesus* are capable of tolerating either fresh or saline conditions, the vast majority are not. Thus the wildlife covered in this book is associated with the middle and lower reaches, the barbel and bream zones that blend together into an interrelated unit. The limit at one end is provided by the grayling zone, while at the other it is the interface with the sea.

Canals

Any fish that inhabit canals must have evolved to fill a niche in a more natural habitat. Logically the dominant canal species should be those of the lower reach, and this is essentially the case, with the roach and the perch as two of the most abundant inhabitants, augmented by a wide range of additional species like bream, tench and the pike. Many of these must have been introduced, but smaller species like the gudgeon and ruffe are more likely to have moved in from associated river systems. For their size and the body of water involved, canals must rank as the most intensely fished waters in Britain. This has many important side-effects for the range of fish species involved. Local introductions and extinctions constantly change the pattern, and it is quite possible to find unusual species like goldfish, *Carassius auratus,* and crucian carp *C. carassius,* thriving temporarily while conditions are favourable. An extreme and celebrated example is the St Helens Canal, where for many years a local factory heated the water and allowed a range of tropical fish, including cyclids, to survive. As soon as the factory ceased pumping, however, the population was quickly exterminated. At Offerton, on the Worcester and Birmingham Canal, full-grown chub occur in the same lock basin as carp: the first a result of natural invasion from the Severn and the second as introduced stock. It is impossible to disentangle the confusing influences that have brought about many similar situations, but at least Britain's canals cannot be accused of predictability.

Canal fishing has one other very significant influence on the water and the species within it. Ground bait is thrown into the water in such quantities that it must have a very marked effect, first by keeping fish stocks at an artificially high level and secondly by feeding scavenging invertebrates. Conversely bio-degrading bacteria take oxygen from the water while decomposing the material, and the water is thereby rendered less attractive to wildlife. Fish are vulnerable to change, are useful indicators of water quality, and are the least often seen of aquatic animals. Whether they can be considered as truly natural is hardly relevant; they are present and their influence is considerable.

7 Semi-aquatic Invertebrates

During high summer, a vast assemblage of insects appears on the riverside, either sitting about on vegetation or buzzing over the water in a frenzied but organised manner. Many of these will be associated with marshland plants rather than with the water itself (see chapter 10), but a large number will have emerged from the water, having spent most of their lives within its constraints.

This group contains some of the most primitive of all insects, having changed little since the Carboniferous Period over 300 million years ago, including dragonflies, mayflies, stoneflies and alderflies. These are characterised by very heavily-veined wings (more advanced insects, like flies, tend to have a much reduced venation) and an aquatic larval stage that develops adult features by a gradual series of skin changes rather than a dramatic transformation. The nymphal state, as this is called, often lasts for more than a year, while the adult stage may only last a few days (or even hours in the case of some mayflies). Even so, the adults are so eye-catching and interesting that they are recognised by a wide section of the public. Dragonflies, and their close relatives the damselflies, are known for their bright colours and grace, mayflies for their delicacy, and the alderflies and stoneflies for their lack of any of these features but with a curious primaeval quality that makes them popular with children.

Their sensitivity to changing water quality has resulted in several of these insects becoming extinct in the last fifty years. *Oxygastra curtisii,* a medium/large dragonfly restricted to lowland rivers, was once found in Hampshire but has not been seen for many years. The mayfly *Ephemera lineata* was once found on the River Thames but the only recent record (1968) is for the Wye, a particularly clean river; the stonefly *Isoperla obscura* was discovered on the River Trent at Nottingham early this century, but has not been seen since heavy pollution for many years made this one of the most lifeless rivers in Europe.

In addition, many formerly common species have become rare and have gained only temporary respite by adapting themselves to the ideal conditions presented by

An ideal habitat for semi-aquatic insects : a disused section of canal, with a good zone of emergent plants

disused canals. The future for several species like the attractive club-tailed dragonfly *Gomphus vulgatissimus* is not very bright.

Apart from the group already mentioned, caddisflies, moths, wasps and true flies also have representatives with aquatic larvae, but in their case there is a distinct pupal stage before the emergence of the adult. These 'holometabolic' insects have the advantage of a straight-forward larval stage that does nothing but eat, but have to progress through a defenceless pupal stage open to attack by any predator that can locate them.

Only the caddisflies (Trichoptera) and the true flies (Diptera) have found water particularly attractive as a habitat. All but one of the British caddis are entirely aquatic as larvae, the exception being a species breeding among the damp waterside moss. They are an important component of the benthic fauna, able to respire by diffusion and therefore able to remain immobile in comparatively deep water. This is not the case with most of the true flies, which need to breathe air from the surface and are often considered to be semi-aquatic because of this. There are a great many of them, however, and over a thousand are associated with freshwater habitats, representing 20 per cent of the total number of species.

Moths and wasps have only made minor inroads, and the few species involved are rarely observed — although the china-mark moths are quite well-known to students of pond-life. Insects with aquatic or semi-aquatic larvae are dependent on two habitats rather than one and this must make them vulnerable, but the dual life has obviously brought success. It has also made them a popular group for study — full of adaptations and transformations, yet with a great deal more to be learned.

Mayflies: Ephemeroptera

A whole mythology has been assembled about mayflies. Their early life is spent under water, but the change from nymph to adult is hazardous and presents them as a vulnerable and tasty food for trout. For this reason they have been the subject of a great many angling stories.

Of the forty-seven species, most are adapted to fast-flowing water, but there are many of the middle reaches and a few specialising in the slow systems of lowland rivers and canals. Since they spend all but a few days of their lives underwater, adaptation in the larval stage is vital; species of upland torrents have compressed streamlined bodies and spend much of their time under submerged stones, while those of sluggish waters are free-swimming or even mud-burrowing. Most feed on algae or detritus.

The change to an aerial life is abrupt; the nymphs either crawl out onto vegetation or rise like corks to the surface of the water. There the skin splits and the fully-flighted insects emerge. The first flight takes them to emergent vegetation or overhanging trees, where they remain for several hours until a second moult takes place. In this extra stage the mayflies are unique among all insects; the shedding of a fine membrane from the whole body — including the wings — reveals an adult a little more colourful than the sub-adult that made the initial flight, but otherwise almost identical. The forewings are large, very finely veined and cross-veined, and the hind wings are either very small or missing altogether (as in the common canal species in the genera *Cloeon* and *Procloeon*). The antennae are very short, while the legs and the tails (usually three) are very long. Sitting on vegetation the wings are held together over the back and the abdomen and the tails arched in an elegant curve, giving mayflies an artistic and graceful appearance.

The obscure life-history has caused some problems for anglers: the sub-adult is called a dun while the adult is known as the spinner. Trout flies are constructed to resemble the insects floating on the water, but sexes and stages vary considerably. *Ephemera danica,* a large species found abundantly on silty streams and chalky rivers, is called the green drake as a dun, while the adult male is the black drake and the female is the grey drake. Once the female has laid her eggs and fallen exhausted to the water she is known as the 'spent gnat'; confusing, but all part of the mythology.

Adult mayflies do not have functional mouthparts, which means that their life is brief, though with seasonal variations from species to species and new generations through the summer, the flight period may be from April to October. The males swarm over the water, rising and falling in a dancing cloud awaiting females. Once a female joins the dance mating takes place in mid air, the pair fluttering down and landing on vegetation, or parting before they reach the water.

Since trout do not naturally occur in lowland stretches of rivers, the mayflies that have adapted themselves to this habitat have attracted much less interest from anglers. *Ephemera vulgata,* which replaced *E. danica* in muddy rivers, is known as the dark mackerel, but the smaller slower-water specialists, like *Caenis horaria* and *Baetis vernus,* remain anonymous though they are an important food-source for coarse fish.

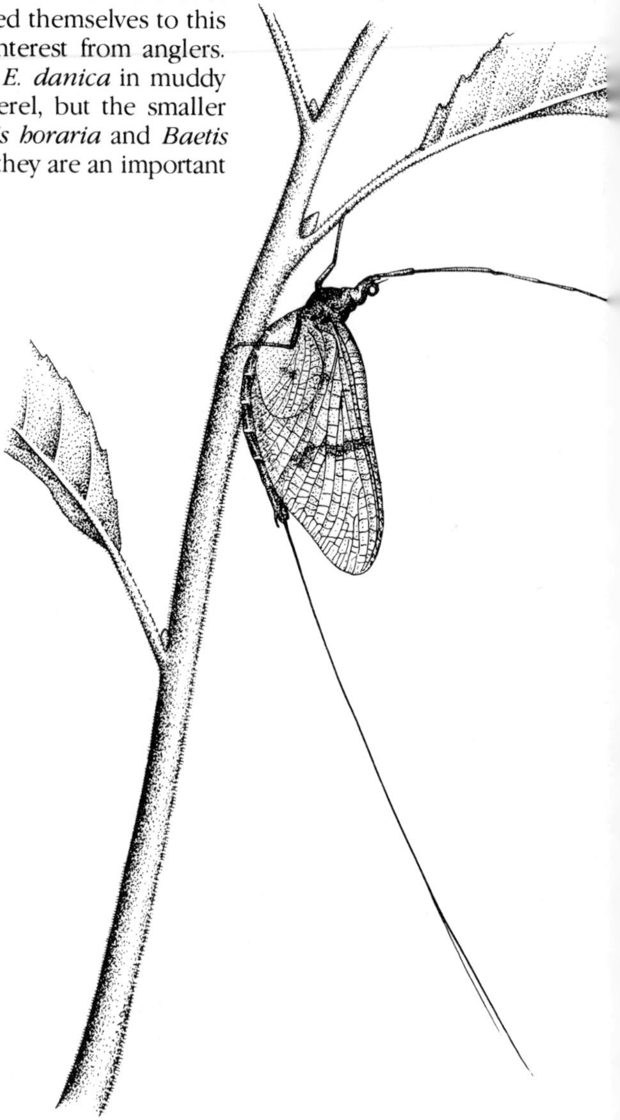

The tails of a mayfly are held together in a graceful curve : **Ephemera vulgata** ***is one of the largest British species and prefers lowland rivers***

Dragonflies: Odonata

This small order of insects comprises only about forty-four British species, separated into the true dragonflies (Anisoptera) and the damselflies (Zygoptera). The latter are a very distinctive group, folding their wings together over their back when at rest and having a long thin abdomen that has earned them the country name 'Devil's darning needles'.

The majority of the dragonflies are associated with standing water: ponds and lakes rather than fast-flowing streams, but sluggish rivers and canals can provide ideal conditions. The most impressive of all the British species is *Anax imperator* (the emperor), a blue hawker with a wingspan of 11cm (4½in) which is characteristic of the clean plant-rich canals of southern England. The *Aeshna* species — particularly *cyanea* (the southern hawker) and *juncea* (the common hawker) — are almost as big and considerably commoner, occuring up to Scotland where

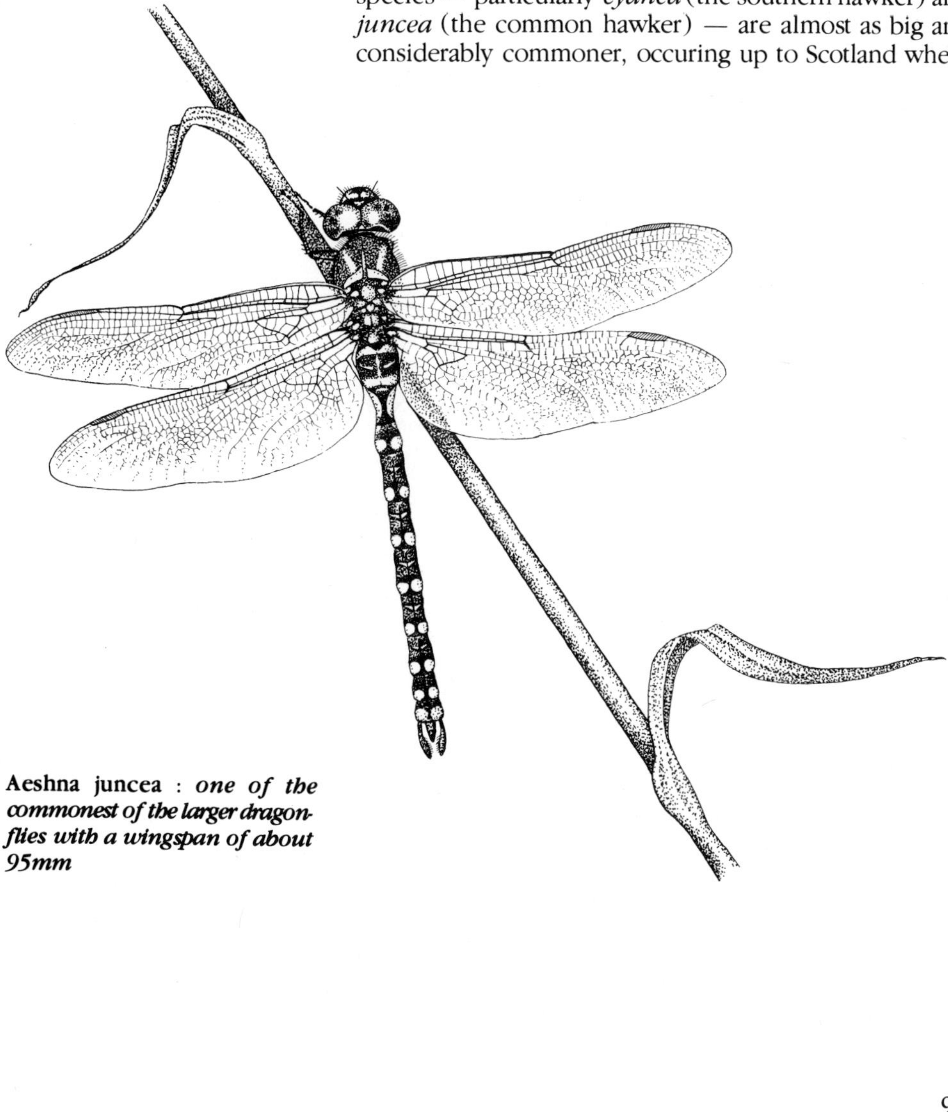

Aeshna juncea : ***one of the commonest of the larger dragonflies with a wingspan of about 95mm***

The blue-tailed damselfly,* Ischnura elegans, *is one of the few species to be found along urban waterways

there are trees close by their selected habitat. In England, *Aeshna grandis* (the brown hawker — a very distinctive species with brown wings) shows a preference for open country rather than woods and is therefore commoner along many of the Midland canals. Most of the large dragonflies are strongly territorial, selecting a beat along a waterway and either viewing this from a high vantage point (for example, the top of a tree) or patrolling at regular intervals to keep out intruders. Many riverside walkers have been disturbed in this way, and to be 'buzzed' by a large hawker is a memorable experience. Some people still think dragonflies can sting, and consider themselves to have been lucky to escape with a warning from an irate *Aeshna*. In fact they have no sting at all, but have a very powerful pair of mandibles and can inflict a sharp bite if they are handled carelessly. This rarely happens, of course; they are the most difficult of all insects to catch, combining superb aerial skills with near-360° vision.

River specialists among the dragonflies are few; *Gomphus vulgatissimus* (the club-tailed dragonfly) is a species particularly associated with the Thames, but most others show a preference for slower and smaller waters. This is also the case with the damselflies, and because they are not territorial they seem far commoner along canals than their larger relatives. The most impressive of the damsels is *Agrion splendens* — the banded damsel, which measures about six and a half centimetres (2½in) across the wings and is larger than some of the dragonflies. It is an exceptionally beautiful creature with a metallic green body, the male having deep blue patches on each of the four wings, while the female has all-green wings; the colour of shaded moss. To see a pair of these *Agrions* fluttering over a richly flowered lowland stream (the preferred habitat) is evocative of stories by H.E. Bates; an age of rural beauty when technology had not become the most important influence on our landscape. Along canals, the smaller damsels are far more abundant. These include *Coenagrion puella* (the azure damsel), *Ischnura elegans* (the blue-tailed damsel) and *Enallagma cyathigerum* (the common blue damsel). All are more or less blue, though the females are much duller and more often mistaken for other species. They are all rather small too, about four centimetres (1½in) across the wings and are very delicate, both in appearance and in their manner of flight. *Ischnura* — with its distinctive blue-tipped tail — is one of the few insects in this group to tolerate polluted water, which is why it is noted on urban canals so long as these have not received too much effluent. *Pyrrhosoma nymphula* (the large red damsel) is widespread too, appearing on a wide variety of waters but not including rivers.

Along many canals, four or five species of damselfly can be seen together, fluttering among the reeds and rarely straying far from the water's edge. *Pyrrhosoma* may be on the wing in early May, but most appear in June, and July is probably the peak period. This is true of the dragonflies too, and many appear at mid-summer and continue on the wing well into the autumn. The long life of adult dragonflies is made possible by a regular intake of fuel, in the form of other insects, consumed in large numbers and often caught on the wing. *Aeshnas* will tackle almost anything that flies, up to and including bumblebees, but their bravery sometimes gets them into trouble and they occasionally fall prey to their prey.

Dragonflies begin their adult life in rather a dull state, and it takes several days after emergence (from the nymphs) before the true colours begin to appear. This period is often spent well away from water, and the return marks sexual maturity, a regular territory being established and a greater period of time each day being spent among trees, particularly by the female which only visits the water to mate and lay eggs. The mating process is rather eccentric, involving the male taking hold of the female by the claspers at the tip of his abdomen, and arching his body to grasp her by the head or neck. A flight in tandem often results, presenting the curious impression of an elongated double dragonfly. The mating process is completed by the female arching the tip of her abdomen to join the base of the male's abdomen, a small pocket of sperms having been deposited there by the male. Why this elaborate procedure has been evolved is difficult to imagine, though it must have significant survival value.

The female lays the eggs either by scattering them while flying over the water or, in the case of the *Aeshnas*, by landing and probing her abdomen among aquatic plants. Damselflies often crawl down into the water to find a suitable site, which must be a hazardous process lasting several minutes.

Stoneflies: Plecoptera

This is a group of very primitive-looking insects that have adapted themselves particularly well to fast-flowing streams. Stoneflies are usually brown in colour and spend most of their adult life hidden among herbage. When they are disturbed their flight is weak and the four wings are opaque and heavily veined, giving them a clouded and clumsy appearance. At rest the wings are folded over the back, often tightly around the body which has earned them the name 'needle flies'. The only species associated with muddy waters is *Nemoura cinerea,* a small and rather insignificant species about a centimetre long which is found along clean canals and rivers where there is a good deal of emergent vegetation. In these conditions it can become abundant but it is rarely noticed because of its retiring habits.

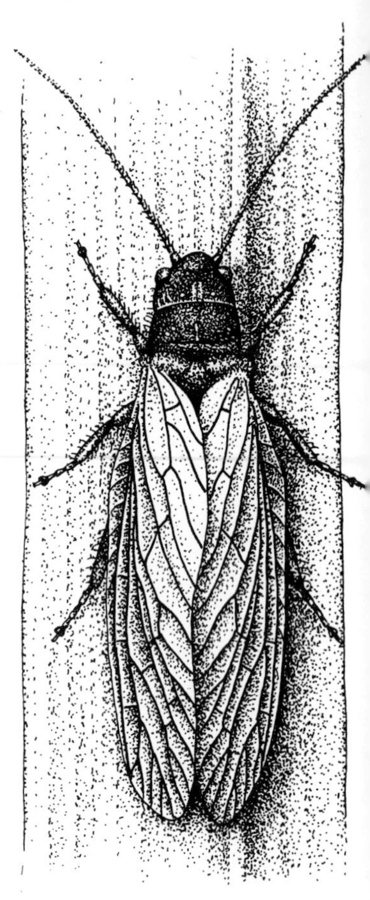

Sialis lutaria ***is the most widespread of the three British alderflies (actual length about 18mm)***

Alderflies: Megaloptera

The egg masses of alderflies are often more noticeable than the adults: laid on emergent reeds or grasses and consisting of several hundred tightly spaced cylindrical brown eggs, they stand out and often give rise to question and comment. The adults are more retiring, spending a great deal of their time sitting on bushes near the waterside, but not necessarily hidden away. Flight and mating usually take place on sunny mornings; the impression of the insects on the wing is that they bear some resemblence to stoneflies, but are much thicker-bodied and even slower.

Until recently, only two British species of alderflies were recognised, *Sialis lutaria* of slow or still waters and *S. fuliginosa* of rather faster rivers. This pairing of species is met with in many other insects, for example the two damselflies *Agrion spendens* and *A. virgo,* but in the case of the alderflies this neat separation by habitat has been confounded by the discovery of a new species, *S. nigripes.* For such obvious and well-known insects as alderflies it seems incredible that *S. nigripes* remained unidentified until 1977, but since then it has been found on widely separated lowland rivers in England, Wales and, most recently, Scotland. It seems likely that it is a rather uncommon species, however, existing at a low density. *S. lutaria* remains the most widespread and most abundant of the three, measuring more than three centimetres across the wings and distinguished from *S. fuliginosa* with difficulty, though the latter is even larger. The flattened brown bodies, sluggish habits and heavily-veined brown wings of both species distinguishes them from most other insects and the method of folding the wings roof-style, apex along the middle, separates them from the stoneflies.

Sialis lutaria flies during May and June and the eggs hatch after only a fortnight: the rest of the life-cycle (up to two years) is spent in the mud of canals or slow rivers, where the larvae consume large numbers of other insects. A tiny parasitic wasp, *Trichogramma semblidis* lays its eggs

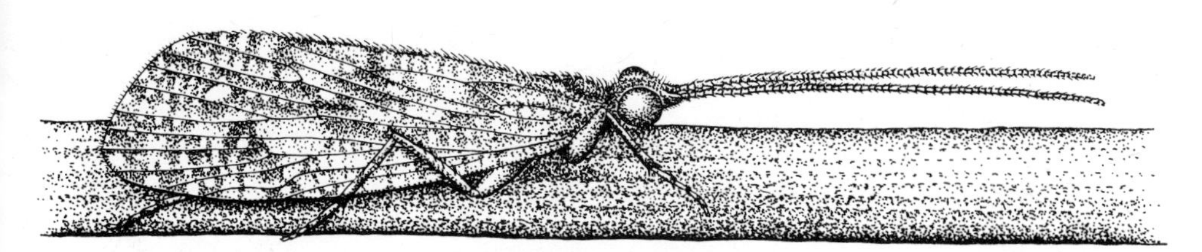

Caddisflies resemble moths but their wings are covered with hairs rather than scales. One of the best-marked is* Phryganea varia. *(actual length, about 16mm)

on *S. lutaria* eggs, each wasp grub finding sufficient food in a single host egg to emerge as an adult in the time it would take for the host larvae to hatch (see separate section on wasps). Only a tiny fraction of the host eggs are attacked in each egg-mass, and the balance of parasite to host and host to its prey is such that the population remain reasonably stable.

Caddisflies: Trichoptera

The common English name, and the angler's variation of 'sedgeflies', disguise the fact that caddis are closely related to moths, having four large hairy wings, a thin body and long antennae. They form a very distinctive group with 189 British species but are very difficult to identify and are usually studied in the larval stage, which the majority spend encased in tubes of sand, stone, sticks or leaf-pieces. The adults are neither colourful nor active by day and sit in inconspicuous places, often tapered along a grass stem with wings folded tent-fashion.

Caddis, like stoneflies, are indicators of clean water and their presence in good numbers is a sign of an unpolluted river. The waterway species are mainly of the kind that use leaf sections to form their larval cases, but there is a particular group called the net-spinners which produce silken webs and have a representative, *Hydropsyche angustipennis,* which will tolerate a poorer quality of water in canals or rivers.

Adult caddisflies have functioning mouthparts; a pair of mandibles designed for chewing rather than sucking, and although these are rarely used they serve to distinguish the group from most moths, while the hairy wings clearly separate them from other insects. The eyes are very prominent and suggest that the caddis are either crepuscular or nocturnal. Anyone living close to a stream or river will recognise them from their regular appearance on kitchen windows on autumn evenings, as they are strongly attracted to light. Few of them draw individual attention; *Mystacides azurea* is an attractive blue-grey colour with very long antennae, and the genus *Phryganea* has large, well-marked representatives like *P. grandis* and *P. varia.* Otherwise they remain incognito to all but the connoisseurs: birds, bats and shrews.

Butterflies and Moths: Lepidoptera

From well over 2,000 species, only about nine moths have

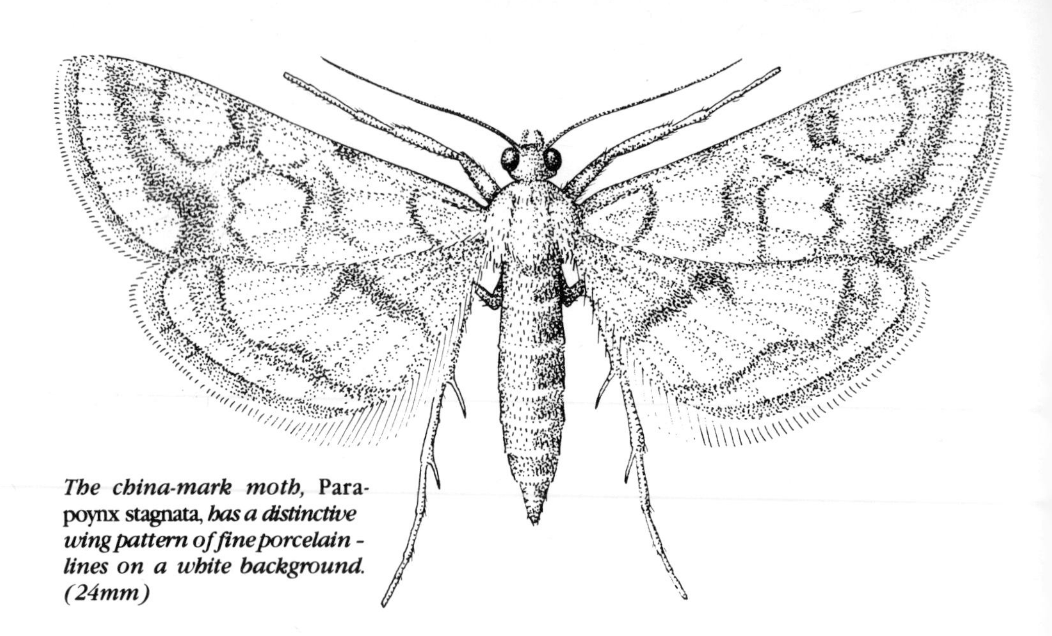

The china-mark moth, Parapoynx stagnata, *has a distinctive wing pattern of fine porcelain-lines on a white background. (24mm)*

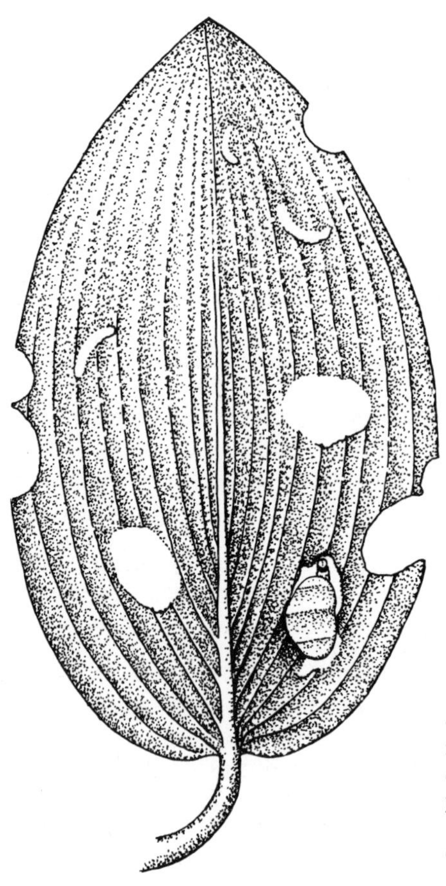

The underside of a floating leaf of the pondweed, Potamogeton natans *with a larva of the brown china-mark moth,* Parapoynx stratiotata *feeding from the safety of a case constructed of leaf sections*

aquatic larvae, and all these belong to the family *Pyralidae* which are small and often rather drab. However,the china-marks are quite well known, firstly because the larvae take slices out of leaves and construct a case in the same way as caddisflies (but not as professional in design), and secondly because the adults are delicate and finely marked. They can be disturbed from marginal vegetation during the day (the main flight is at dusk) and are particularly common along old canals. *Parapoynx stratiotata,* the brown china-mark, abounds wherever pond weed occurs, but like its attractive relatives *P. stagnata* and *Nymphula nymphaeata* it is not restricted to this foodplant; bur-reed or frog-bit will serve just as well. The other common species of china-mark is *Cataclysta lemnata,* confined to duckweed and still very common where the water flow is restricted, but dissimilar to the others in that it does not have any fine lines to its wings. The male is mainly white and the female brown, with a short string of black pearl markings on its hind-wings.

A less well known group of moths mines into *Phragmites* reed stems, the larvae floating on small leaf-rafts from one plant to another as these are consumed. The main species are *Schoenobius forficella* and *Donacaula mucronellus* — both are small, brown and easily confused with the many other small brown moths in similar habitats.

The most dramatic of the aquatic species is *Acentria nivea,* not because it is particularly large (in fact it only measures a few centimetres across) but because the female never appears above the water and has wings reduced to tiny stumps. The male meets her at the surface, where a brief mating takes place before she disappears down to lay her eggs on pondweed. Surprisingly it is a very easy species to spot; the male is pure white in colour and engages itself in long meandering flights a few centimetres above the water. Again, disused canals present the optimum habitat, but *Acentria* usually stays well out over the water and is best identified at a distance.

Wasps, Bees, etc: Hymenoptera

Only a fraction of this very large group is associated with water and those with a place in the economy of aquatic biotopes are all wasps and act as parasites. Since they kill their hosts they are not parasitic in the true sense, however, and are often referred to as parasitoids. The tiny wasp *Mestocharis bimacularis* is a typical example. It lays its eggs on water-beetle eggs, but can only do this when the water level has dropped sufficiently to expose these temporarily. An advance is exhibited by a Mymarid called *Caraphractus cinctus,* which also lays its eggs on those of *Dytiscus* water beetles but spends its whole life in the water, using its wings as underwater paddles. The size of *Dytiscus* beetles is doubtless the reason why their eggs are so often the subject of parasitism, but Mymarid and Trichiograminid wasps also attack the eggs of dragonflies and alderflies that have been laid on emergent vegetation.

The only wasp much larger than a millimetre to descend into the water is the Ichneumon *Agriotypus armatus*, which is parasitic on caddisflies and dives below the surface to lodge its eggs in the prepupae (the dormant larval stage) of its selected host. The larva of the wasp eats the caddis internally and emerges as an adult in the autumn, but remains beneath the surface in the old caddis case until the spring. Like most other parasitic wasps, however, *Agriotypus* is still quite small, slim, fast in flight and rarely noticed in the adult stage. It is thought to be very rare, but could easily be overlooked and may yet be discovered on unexpected waterways.

Flies: Diptera

Midges, mosquitoes and craneflies account for a very high percentage of all insects with aquatic larvae, and the majority of those associated with flowing water are both harmless (ie non-biting) and unobtrusive.

The biggest of the British flies is the 'daddy long legs' *Acutipula maxima*, with a wingspan of 65mm (2½in); there are over a hundred other Tipulids ('daddy long legs'), characteristic of the shallow edges of rivers and canals as well as drainage ditches, ponds and streams. The larvae are grey 'leather jackets', similar to those found in lawns, but with minor adaptations to cope with being underwater for most of the time. The adults, with their clattering wings and trailing jumble of legs, are not the most popular of insects and could hardly be described as unobtrusive, but at least they do not bite!

Mosquitoes (Culicidae) are more readily associated with stagnant water, but urban canals can harbour large populations of such species as *Culex pipiens*. Only a handful of the fifty species suck human blood, and even then it is only the females that are responsible. Other closely related families are the gnats, *Dixidae* and *Chaoboridae* which include several important aquatic predators — for example the phantom midge *Chaoborus crystallinus* — and the true midges, Chironomidae. These do not bite but are inclined to settle quietly on moving targets and get themselves swatted in mistake for mosquitoes. The well-known 'blood-worms' are the larvae of *Chironomus* midges; many are again found in stagnant water where there is little oxygen, but a large percentage of the 380 British species dominate slow-flowing systems and provide a staple food for fishes of all kinds. The other important group of midge-related flies is the *Simuliidae*, called black-flies but rather insignificant until they bite. These are well-known insects because the larvae feed by filtering food from fast-flowing water, and because certain

▷ ***One of the most abundant of waterside birds, the moorhen,* Gallinula chloropus, *which feeds on aquatic and semi-aquatic plants***

▷ ***A kingfisher,* Alcedo atthis, *ready to feed its family deep in the riverbank. If it were going to eat the fish itself it would be holding it head first***

A reed warbler,* Acrocephalus scirpaceus, *feeding a well grown cuckoo* Cuculus canorus. *Reed swamps are always attractive to cuckoos because of the high population of potential hosts

◁

foreign species can transmit disease and cause a great deal of damage to livestock.

Not all flies are drab, and some of the semi-aquatic species are very handsome. Soldierflies, belonging to the family *Stratiomyidae,* are all elegant and harmless, the adults being attracted to waterside flowers. The biggest and most dramatic, is *Stratyomys potamida* which is fairly common in the south of England. Other smaller species, all brightly coloured, are found throughout Britain.

Hoverflies, belonging to the family Syrphidae, include several 'aquatic' species like *Eristalis nemorum,* known in the larval stage as rat-tailed maggots and found in the sort of place that not even mosquitoes would find hospitable. The adults love sunny glades, however, and are known for their habit of hanging motionless in mid-air, always avoiding outstretched hands but returning to the same few centimetres of air-space. They look rather like bees, and their country name of drone-flies reflects their similarity to the large-eyed male hive-bee of late summer.

Apart from the families Empididae and Dolichopodidae — numerous but insignificant to the general naturalist — the only other notable family of flies associated with water is the Tabanidae. This includes the bulkiest and most frightening of all our Diptera, known as clegs or horseflies. Very impressive beasts, fast and silent with iridescent eyes and a bite to bring tears to the eyes of the most stoical entomologist. *Tabanus sudeticus* is the largest of the semi-aquatic species: more typical is *Chrysops relictus* which is sometimes called the thunder fly because it is usually in evidence on humid close days. It is a very beautiful insect with banded wings and glowing purple eyes; beautiful but painful, and a bane to any country walk.

8 *Birds, Reptiles, Amphibians and Mammals*

The vertebrate animals include the best-known but most elusive of our waterside wildlife. Most of the animals are nocturnal, while the reptiles and amphibians are shy or simply go unnoticed. The majority of birds seen along waterways are only there temporarily and most of the characteristic species — like the kingfisher — seem to be remarkably difficult to find.

A recent checklist of animals occurring in freshwater includes eight mammals, eight amphibians, two snakes and 205 birds. At first glance, this suggests that birds figure prominently in the economy of water habitats, but the vast majority are either better suited to marsh or estuary, or are arbitrary visitors making use of a temporary food source. The species that depend on waterways include representatives from several families with a wide variety of individual requirements. Those that feed on aquatic plants and animals often remain throughout the year, but the insectivores have to migrate away when the summer abundance has waned. Thus the picture changes from season to season, and even 'resident' species may be obliged to move according to the local conditions.

The ability of birds to react rapidly in adversity gives them a clear advantage over other animals, and makes them early indicators when anything happens further down the food chain. Climatic changes, increases in pollution or subtle variations in amenity-use can affect birds in spectacular ways, yet their adaptability enables them to cope with any but the most serious problems. The more sedentary the animal, therefore, the more dependent it is on local conditions, so a decline in a frog population is rather more significant than the disappearance of herons from a particular stretch of water.

Most warblers move to North Africa for the winter, and this process is the controlling factor in their mortality. 77 per cent of the total British population of whitethroats disappeared between 1968 and 1969. It is assumed that a drought in the Sahel area below the Sahara was responsible for this, but although the loss of 8 million birds of one species may be unusual, the mortality for birds like the sedge warbler and blackcap must be appallingly high even in a normal year.

Those birds that remain in Britain for the winter may be partial migrants, moving across the country as conditions dictate. This mobility is shared by the larger waterside mammals, the otter and the mink, and perhaps even the water shrew. The reptiles and amphibians rely on hibernation to take them through the winter, and in their case it is this process that leads to the greatest mortality. Winter therefore is a testing time for all vertebrates, and the spring populations are very much lower than those of the autumn.

Because of their shyness, observation of birds and mammals is always difficult: the main problem is to find them, and once this is accomplished identification (unlike with the invertebrates) is usually straightforward. The most important factor is the time of day, since most vertebrates are at their most active at dawn and dusk. Those that are not specifically adapted to this twice-daily rhythm have adopted it to avoid disturbance by man. It follows that the easiest animals to see are those like the water vole: common, active during the day and tolerant of man, while the most elusive are those that, like the otter, are semi-nocturnal, rare and intolerant of any disturbance. Most animals fall somewhere between these two extremes, and given a reasonable amount of luck and a good deal of footwork most of the characteristic waterway species can be seen in a single year on one river or canal system.

BIRDS

Of the twenty-five species included here, the most often seen are the moorhen and the mallard, and this is a fair indication of their abundance. On the water-edge the pied wagtail, reed bunting and sedge warbler are the most widespread, but there are those with a northern bias like the sand martin and, similarly, southern specialities like the reed warbler.

Upland breeding birds often transfer to lower altitudes for the winter, and this happens a good deal in water-dependent species. The dynamics of food availability and bird movements are much less pronounced in species like the dabchick and moorhen that have taken to canals and are only forced out when ice makes it impossible for them to remain.

It is difficult sometimes to decide which are truly waterway birds and which come into contact with rivers and canals only infrequently. The yellow wagtail is often thought of as a riparian species yet it feeds and nests in damp meadows and is rarely seen on the water-side. For this reason it is considered in the next chapter. Rare river-breeders like the Cetti's warbler are also excluded, as are less common migrants like the greenshank and green sandpiper

Canals and drainage channels provide ideal conditions for the dabchick, **Tachybaptus ruficollis,** ***particularly when there is abundant emergent vegetation***

Dabchick *Tachybaptus ruficollis*
Otherwise known as the little grebe, this is a small dumpy bird with hardly any tail and a fluffy stern held high out of the water. It is often mistaken for a duckling, but its short pointed beak is distinctive. During the summer it is very secretive, choosing a section of slow-flowing river or canal where there is a lot of vegetation and where it can disappear when alarmed. The nest is a floating island of weeds, usually hidden among reeds or half-submerged sallows. It dives a good deal, feeding on small fish and invertebrates like water-beetles and damselfly nymphs, and often resurfaces out of sight. During the summer it is more often heard than seen, the call being a very loud whinny or trill which carries a surprisingly long way. During the winter, dabchicks change plumage from dark grey/brown with reddish cheeks to paler grey with white cheeks and neck. Some move to gravel pits or reservoirs where they may congregate in some numbers, but the majority, especially the adults, remain on their favourite stretches of water.

Great Crested Grebe *Podiceps cristatus*
This is twice the size of the dabchick, with a long pointed

beak adapted for catching fish. During the spring and summer the great crested grebe has cheek tufts and crests, but during the winter these are almost completely lost. There were only thirty-two pairs nesting in Britain in 1860, but today there are probably over 6,000 nesting mainly on lakes and gravel pits. The increase in the population is probably responsible for the recent expansion onto slow-flowing rivers in the south and east of England.

Outside the breeding season, these grebes are often seen on waterways, but they still prefer larger lakes and reservoirs.

Heron *Ardea cinerea*

An unmistakeable bird of the waterside, but nesting some distance away in tall trees. Herons will eat almost anything that moves, from young rabbits to frogs, crayfish and eels. When eels are eaten they writhe around and cover the heron's head with slime; this is removed by special 'powder down' which acts rather like talcum powder and is preened into the feathers on the face and neck.

The heron, **Ardea cinerea,** ***is an adaptable hunter but suffers dramatic declines during severe winter weather***

Although herons can be found in almost any wetland areas, they are characteristic of wide, slow-flowing rivers and marshes and do not often visit canals. The best time to look for them is at dawn and dusk, when they are in search of food and have not been disturbed.

Cormorant *Phalactrocorax carbo*
Cormorants nest on the coast and the majority stay there, but many wide lowland rivers are visited during stormy weather, and there are several places (for example, on the Old Bedford River at Welney in Norfolk) where large winter roosts may develop. When on the water, cormorants are very distinctive: large, low in the water, with snake-like necks and dagger-shaped beaks.

Mute Swan *Cygnus olor*
The mute swan is one of the heaviest and most powerful of all flying birds, and certainly can look after itself. The stories of swans breaking peoples' legs are an exaggeration, however, though a 'cob' (male), fluffed up and hissing, commands respect and is best left unprovoked if it has a family nearby. The nest is a huge mound of reed, built along a disused canal or quiet backwater, and when the young cygnets are a few weeks old they are often taken to more extensive waters where there is more aquatic vegetation. Young birds and other non-breeders congregate into flocks, and these became quite famous at such places as Stratford upon Avon and on the Trent at Nottingham. Their drastic decline during the late 1970s has been blamed on lead poisoning, resulting from birds swallowing discarded fishing weights in mistake for grit.

Many swans die as a result of flying into overhead cables, but some are illegally shot by farmers when they stray onto agricultural land to feed. Disturbance to the nesting birds may cause some pairs to fail, and there seems to have been an overall decrease in recent years. In spite of this, most lowland rivers and canals have their regular pairs of mute swans, which mate for life and may live for twenty years.

Canada Goose *Branta canadensis*
A very large goose, with a long black neck and white cheeks, introduced from America 300 years ago and now quite at home here. The Canada stays here all year round (most other geese migrate north for the summer), nesting on gravel pits and ornamental lakes (where it can cause problems) and visiting riverside grassland to feed.

Shelduck *Tadorna tadorna*
An estuarine species which is sometimes seen on river systems inland. Shelduck, with their distinctive pied plumage, nest in burrows and have taken to breeding some distance inland — for example on the River Nene in

Tidal rivers (in this case the River Nene near Guyhirn in Cambridgeshire) attract waders and shelducks which are drawn away from their normal coastal haunts

Cambridgeshire. They are far more characteristic of coasts however, feeding on tiny *Hydrobia* snails on extensive mudflats to be found on such estuaries as the Thames and the Severn.

Teal *Anas crecca*
A small, neat duck, shy and less inclined to appear on open water than the mallard. During the breeding season, it prefers upland pools, but may also nest on lowland lakes and rivers with plenty of cover.

Outside the breeding season, teal usually appear in small groups, dabbling for food in very shallow water at dawn and dusk, and roosting on open water away from human disturbance.

Mallard *Anas platyrhynchos*
The most successful of all water birds, despite the fact that it is heavily shot and has been an important human food source for several hundred years.

Mallards occur in all kinds of wetlands, and on every

Mallard, Canada goose and moorhen tracks to and from the River Clun in Shropshire

kind of waterway. The nest is usually in deep cover and may be some way from the water. Those that nest too early (and in southern England this may be in early February) usually lose their brood to foxes, stoats or crows, which can see the nest before the vegetation has had chance to grow. Most ducklings appear in April and May with additional broods through the summer and early autumn. At this time the adult drakes lose their attractive colours and adopt an 'eclipse' plumage while new wing-feathers are grown. They become as shy as the brooding females at this time and it often seems that whole stretches of river have been deserted.

Urban waterways often have large congregations of semi-tame mallards which feed on bread and add interest to what otherwise might be rather depressing riverside walks.

Tufted Duck *Aythya fuligula*
This is a diving duck, feeding at a depth of 3-5m on a wide range of aquatic life (but mainly molluscs and crustacea). During the breeding season, the female often selects a nesting-site well away from water, and although this is not

a species associated with rivers and canals, the rapid increase that has occurred in Britain in the past thirty years (from a breeding population of none in 1949 to 5,000 in 1980) may be reflected in a corresponding increase in their selection of quiet waterways close to lakes or reservoirs.

Goosander *Mergus merganser*
One of our most beautiful duck, nesting on fast flowing upland rivers and moving to lowlands — but especially reservoirs — for the winter. Numbers seem to be increasing despite persecution (it has the unfortunate habit of being too efficient at catching young salmon and is not protected by law). Small groups appear on most river systems at some time or other, but often as temporary visitors during hard winter weather.

Water Rail *Rallus aquaticus*
Not an easy bird to locate! It produces a curious pig-like squeal which is often the only indication that this species is present in an area, but at dawn and dusk it can sometimes be seen away from the deep reed-cover that it usually prefers. The nest is made among reeds (Phragmites) or reedmace (Typha) or on quiet stretches of slow rivers or disused canals. During the winter, juvenile birds disperse widely, and in hard weather they can turn up almost anywhere where there is a tangle of wet vegetation: drains, ditches and puddles.

Moorhen *Gallinula chloropus*
The long breeding season — from March to August — and the ability to lay replacement clutches allow moorhen to counter a high rate of predation and maintain a very high population. Canals are particularly favoured, but slow-flowing rivers, ponds and drainage ditches are used almost as frequently. Moorhens are aggressive and territorial and will not tolerate intruders of the same species; consequently they never appear in large numbers and remain in the same area throughout the year.

The nest is built on a platform of dead vegetation and is often quite conspicuous, which is probably why so many of the early broods are destroyed. When heavy rains are due, the platform is often raised by the addition of several inches of fresh reeds and grasses, but this often takes place in desperate panic rather than as an orderly premonition. Moorhen eggs have always been a food for country folk, and this tradition is certainly continued in many countries, whether it is legal or not. The main natural enemies are the fox, stoat and weasel, and many of the black downy-young fall prey to pike.

Coot *Fulica atra*
Although usually associated with more open waters, the

coot has become an important breeding species on rivers and canals of south-east England. The main requirement is for a good supply of vegetation which forms most of the diet. The coot is the clumsiest of all diving birds and seems to expend a great deal of energy getting itself underwater. Weed is plucked from a depth of a metre or less and eaten on the surface, or fed to the young which often chase the adults in order to be first in the queue. The white beak and head-patch and larger size distinguish it from the moorhen; otherwise the two are rather similar and often confused.

Oystercatcher *Haematopus ostralegus*
During the winter and throughout most of its breeding range, the oystercatcher is a coastal species but since 1900 the breeding range has extended to include the courses of many northern rivers. The reason for this sudden change has been attributed to the new habit of feeding the young on earthworms rather than molluscs. Certainly inland nesting is proving very successful and is likely to spread further to the south and east. Oystercatchers are unmistakable, with a contrasting black and white plumage and bright red beak.

Redshank *Tringa totanus*
In the south the wild neurotic cry of the redshank is associated with the coast, but in the north this is a typical riverside wader, breeding in upland marshland and moving down to rivers later in the year.

The diet consists of abundant mud-living invertebrates, which are probed for or picked up from the surface. The principal food on estuaries is the small snail *Hydrobia* and the crustacean *Corophium,* but since these are relatively uncommon in freshwater alternatives must include a large number of worms and fly grubs.

In flight the best identification feature is the white wing-flash; the long orange legs may be covered with mud!

Common Sandpiper *Tringa hypoleucos*
A wading-bird with rather short legs, characterised by its habit of bobbing in a nervous manner as it searches for invertebrates along the sides of rivers and pools. Nests along upland (stony) streams, but appears commonly along lowland rivers during the autumn and, to a lesser extent, the winter.

Kingfisher *Alcedo atthis*
No other bird excites attention on a riverside walk as much as a kingfisher, but it is an achievement to see anything more than a blurred flash of blue as a bird dashes past, low over the water and intent on avoiding any contact with the human race.

Footprints of the common sandpiper, Tringa hypoleucos, *on a tidal riverbank*

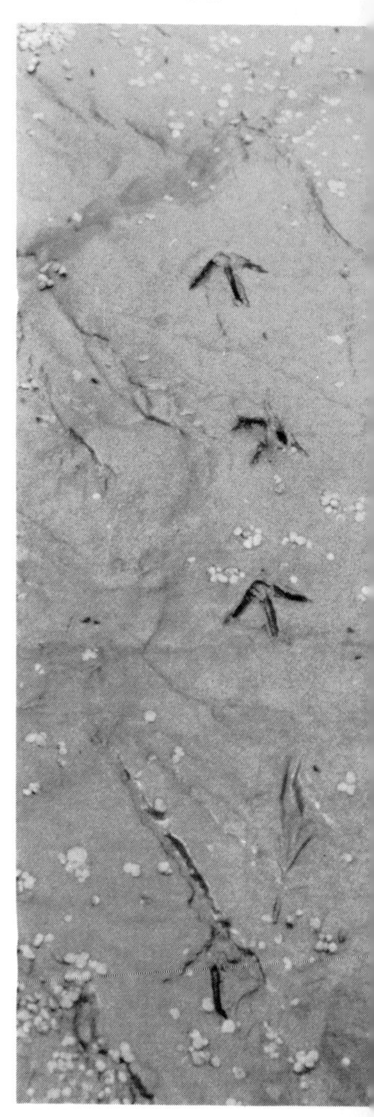

The short wings, tail and legs of the kingfisher, Alcedo atthis, ***contrast with its large head and powerful beak. When relaxed it is a remarkably dumpy bird but its electric colours compensate for any lack of grace***

Rivers are by far the best habitat, mainly because steep eroded banks provide ideal breeding sites: holes are excavated to a depth of about a metre (3ft) and the nest cavity itself is carpeted in an untidy layer of decaying fish bones.

Kingfishers prefer slow-flowing waters but will often nest in woodland or along dried-out water-courses, so long as there is a suitable bank within easy reach of a supply of minnows or other coarse fish. The controlling influence on kingfisher numbers is the severity of winter frosts: after very hard winters the population is decimated, but recovery is always fast. Broods are large and the young birds are capable of very wide dispersal, taking them far and wide in all directions.

Pollution may have affected the number of waterways suitable for colonisation, and the absence of breeding pairs around Bradford and Leeds remains a testament to the state of canals and rivers in that area.

Sand Martin *Riparia riparia*
Feeding on the wing on midges, caddis flies and other freshwater insects, the sand martin is a classic example of a bird suited to river conditions. Nesting is colonial and is dependent on the availability of vertical banks close to the water. In the south of England this usually takes the form of gravel pits, but in the north and west river banks provide ideal natural conditions. The nesting tunnel, excavated in painstaking fashion, extends about a metre (3ft) but, unlike the kingfisher the sand martin lines its nest with feathers or other material caught in flight over water.

Smaller than the swallow (to which it is closely related) the sand martin has only a short forked tail and is distinctly brown above and white below, with a dark band across the breast. As with most migrants, numbers can fall drastically following bad weather or drought in the 'fueling up' stage of their journey from North Africa.

Grey Wagtail *Motacilla cinerea*
This is one of the most elegant birds of the river, particularly the male which during the breeding season has a black bib, grey head and back, and lemon-yellow underparts.

Nesting is always near fast-running water, and upland streams are the ideal habitat. In lowland areas weirs and waterfalls provide the only suitable territories, so the grey wagtail in summer is quite a rare sight in the south and east. Canals are virtually ignored.

During the autumn there is a dispersal downstream, and the winter is spent along river banks that can provide a good supply of food. Grey wagtails spend most of their time on the very edge of the water, catching damselfly nymphs, water-beetles or other aquatic insects. Cold winters can send them on to the coast and cause population crashes almost as severe as in the kingfisher.

Pied Wagtail *Motacilla alba*
Unlike the grey wagtail this species seems to prefer lowland rivers and is also far more common. Its food is more varied too and it is not confined to water habitats. Despite this, its alternative name of water wagtail is apt for it shows a preference for pools, sewage works and water-meadows where insect larvae are abundant.

Dipper *Cinclus cinclus*
Another bird that breeds along fast-flowing rivers or waterfalls and moves downstream during the autumn, but not quite so widespread as the grey wagtail. It is also completely unlike the wagtail in that it resembles a great wren — short and stocky — and feeds by 'walking' submerged in the water to find mayfly nymphs and other invertebrates. It can swim extremely well for such an

unsuitably-shaped bird, but it is usually seen in low flight over the water when its shape and colour (brown with a white breast) are distinctive.

Sedge Warbler *Acrocephalus schoenobaenus*
The song of the sedge warbler — a chattered jumble of repeated notes, sometimes incorporating odd phrases from other birds — is a common sound on canals and slow-flowing rivers where there is ample vegetation.

Small birds are always difficult to identify, but the sedge warbler, **Acrocephalus schoenobaenus,** ***has a distinctive eyestripe and often appears in full view on waterside sallows***

Sallow scrub is the ideal habitat, especially where this is growing among *Phragmites* or *Glyceria* reeds.

Less secretive than the other warblers, the song is continued in short flights when the orange rump and streaked plumage can be seen. At rest or searching for insects among foliage, the pale eyestripe is very distinctive and makes identification quite easy. Sedge warblers arrive in Britain in early April and the majority have left by the middle of September.

Reed Warbler *Acrocephalus scirpaceus*
Arriving a fortnight later from its winter home in North Africa, this bird shares water habitats with the sedge

warbler but it is restricted to areas where *Phragmites* reed is dominant and this makes it far more of a lowland species. It is more secretive too, rarely flying out from amongst reeds, where it feeds on flies and other insects. Since it is less often seen — and only then it is a 'little brown job' flitting low down from stem to stem — it is best located by its powerful song, which resembles that of the sedge warbler but is much less garrulous. Reed warblers often play host to cuckoos, which is why reed-beds are always good places to hear early cuckoos in late March and early April.

Reed Bunting *Emberiza schoeniclus*
A bird that seems to find canals particularly to its liking, where it selects the same habitat conditions as the sedge warbler but it is twice as common. Unlike the warblers, buntings are seed-eaters and this enables them to avoid the hazards of migration, though there is often a dispersal to the south in the autumn to find easier feeding conditions.

It is characteristic of waterways, but can often survive quite well on farmland, and seems to have adapted itself to drier conditions as the number of wetland sites has decreased. The male has very distinctive plumage: head and neck black with a moustache-stripe and white collar. The female is more difficult to identify, resembling a streaky sparrow but with white outer-tail feathers.

AMPHIBIANS AND REPTILES

Over most of Britain only one frog and one toad are found, but several other species are included on the British list. Of these there are two introductions, the edible frog (*Rana esculenta*) and the marsh frog (*Rana ridibunda*). Neither of these has spread very far, and the edible frog in particular has a long history of introduction and subsequent extinction. The marsh frog maintains itself quite well on the River Rother and the old Royal Military Canal on the borders of Kent and Sussex, and has shown slow but steady expansion since its introduction in 1935.

The natterjack toad is certainly a native, but is restricted to marsh sites near the coast and is not a creature of waterways. This leaves the traditional two species, the common toad ('tadie' in Anglo Saxon) and the common frog.

There are three British newts, all found in slow or still water and all considered here, though the palmate is more of an upland speciality.

Of the reptiles, the grass snake is associated closely with water, but apart from the smooth snake (which is extremely rare and is not included) the others all keep away from aquatic habitats.

All of the amphibians are predators and play an important part in the ecology of freshwater. They are also very popular with children and teachers and it has been argued that frogs are becoming rare because of their popularity as specimens for dissection. This seems unlikely, but it is certainly true that every year tons of both spawn and tadpoles finds their way into the sewage system, having been flushed out of harms way by the parents of frustrated young zoologists.

Warty (Crested) Newt *Triturus cristatus*

All newts hibernate away from water, but while the other two spend the greater part of their lives on dry land, the warty newt is not in a hurry to leave the breeding sites once egg-laying has finished. It is the largest species, an average male measuring about 14cm (5½in), though the female is smaller. During the spring the male has a tall saw-toothed crest along its back and tail and the underside is bright orange with black spots. The crest disappears as the breeding season is passed and the female never develops one at all. While the name crested newt is inappropriate then, the alternative of warty newt describes the defence mechanism well. When it is attacked by predators the glands produce a very unpleasant-tasting substance: even a grass snake will spit it out and, presumably, remember not to try one again.

Commonest in the south-east and apparently preferring deep ponds to canals, the warty newt is not often met with on waterways and is quite local in those areas where it does occur.

Smooth Newt *Triturus vulgaris*

This is the common newt over lowland Britain and is to be found on most clean canals. It is in the water as an adult for three or four months from early spring onwards, but by August most will have transferred to a life among waterside vegetation, hidden under tree-stumps and stones. When on land the smooth skin becomes velvety and the colour changes from olive green with an orange-red belly to a rather dirty brown.

The female lays several hundred eggs in April and May, each singly and hidden among aquatic vegetation. The tadpoles that develop have external gills and are pale brown in colour. They remain in the water until October, by which time the gills have gone and the legs have fully developed.

Newts make poor pets and do not take to life in an aquarium — particularly if they cannot get out of the water.

Palmate Newt *Triturus helveticus*

To some extent this species replaces the smooth newt in the north and west, but there is a wide overlap and the palmate newt is often found in canals in the Midlands and Eastern England. It is the smallest British species, measuring about 8cm (3in), but the male in particular is very attractive. During the breeding season it can be distinguished from the smooth newt by the fact that the hind feet are webbed and the tail has a short thread at its tip.

Common Toad *Bufo bufo*

The eyes of a toad are an attractive copper colour — otherwise it is a creature with few obvious endearing qualities. And yet it holds sufficient interest as a species to be a popular subject for study, and as a creature of character to be the anti-hero in *The Wind in the Willows.*

The toad spends most of its life under logs or stones, emerging at night to feed on insects or other land invertebrates. Hedgerows along canals are ideal, providing breeding sites close at hand without a dangerous migration to and from water. Once in the water, spawning takes place quite quickly, usually in March, sometimes when there is still ice on the surface.

Canals are particularly suitable for toads because of their depth; frogs prefer shallower water and the two are rarely found together, either as adults or tadpoles.

Male toads are much more common than females, and when large congregations occur at mating time the females are easily distinguished — by their larger size and by the fact that they usually have an entourage of males, with one (or sometimes more) clinging to their backs in 'amplexus'. In this way the long string of spawn is laid and fertilised.

Back out of the water, toads resume their nocturnal life-style until hibernation in October or early November.

The warty skin, with its distasteful secretion, protects the toad from many enemies, but it is loosely-fitting and some birds — like the heron — have developed the habit of turning over their prey and either gutting or shaking them until the contents are thrown out.

Migrations of toads are quite common — either adults on their way from spawning, or young ones on land for the first time. This often happens during heavy rain, which has led to the curious myth of showers of toads appearing, raining down from the skies.

Common Frog *Rana temporaria*

The life cycle of the common frog is very similar to that of the toad. It shows a preference for ponds, however, and is not so readily associated with rivers or canals unless these are shallow. Life out of the water is spent in herbage without any special hiding place, and the diet is less random than in the toad. A great many snails and slugs are eaten, but in their turn frogs themselves are eaten by many waterside predators. Most hibernate underground, but

Shallow canals provide suitable spawning grounds for the common frog* Rana temporaria, *but rivers are usually too deep and swift-flowing

some bury themselves in pond-mud and remain underwater from October until February.

Frogs come in most colours and sizes: they can be brown, green, yellow or almost red, but are rarely the dark brown/grey of toads. The skin is smooth without any warts, and there is a characteristic black patch behind the eyes. Because they are diurnal they are often seen on country walks, but there has certainly been a decrease in numbers corresponding with the drainage of farm ponds and ditches.

Grass Snake *Natrix natrix*

The distribution of this large, attractive snake does not extend into Scotland. It is mainly a lowland species with a distinct preference for wet places, and canals and slow rivers provide very suitable habitats.

The grass snake swims well, and obtains much of its prey from the water; frogs are the most important item on the menu, but it will also take fish, newts or large insects.

Like other reptiles, it is active only by day, but is rarely seen since it is very sensitive to vibration and will quickly disappear into undergrowth when approached. Although it has no venom, the defence when provoked or cornered is to hiss and strike, with the mouth closed, or to eject a vile-smelling liquid from the vent. Mating is in the spring, egg-laying in the middle of summer; this takes place in a 'nest' hollowed out among leaf debris, particularly in places where the temperature is higher due to the composting action of the leaf mould.

Grass snakes are easily identified by their yellow collar-marking; they are commoner than most people think and are one of the most beautiful and interesting of the waterway predators.

MAMMALS

Of the semi-aquatic mammals found in Britain only three are truly native, the rest being either marine (the seals), extinct (the beaver) or introduced (mink, coypu and the musk rat). Animals that are adapted to waterways have good fur, so it is hardly surprising that all the introductions have been to supply the fur trade in the 'twenties and 'thirties.

Our three indigenous water animals include the otter, the water shrew and the water vole, and all three have become part of the river mythology. The otter is adapted to its way of life in a number of ways — in particular it has webbing between the toes and a strong muscular tail. The water shrew has modifications along the same lines, but in its case these extend only to fringes of hair around the hind toes and under the tail. The water vole has no special adaptations at all and is similar in most respects to other members of its family. Since it feeds on vegetation it hardly needs to travel at speed, and is one of the most easily-seen mammals of either aquatic or terrestrial habitats.

None of our waterside mammals strays very far away from home, but all are capable of cross-country migration if the need arises.

Water Shrew *Neomys fodiens*
For its size, the most ferocious animal to inhabit waterways, tackling prey larger than itself with surprising tenacity. Although quite common along cleaner rivers and canals, the water shrew is one of the most difficult animals to see since it is small, often hidden under cover, and is less approachable than the water vole. It inhabits a tunnel system on the water margin and spends most of its time either on the water's edge or swimming in search of prey (invertebrates, fish, frogs, etc).

Easily separated from the other shrews by its larger size and its colour — black with (usually) a silvery white underside — the water shrew is active at any time of the day. Its main enemies are larger mammals like the weasel, stoat and the mink, but the pike probably account for quite a number, especially those that are young and inexperienced.

Beaver *Castor fiber*
Once very common on the Fens and along the river systems of Wales (for example, the Teifi), but, according to Pennant, extinct since the twelfth century. Attempts have been made to reintroduce the beaver, but the chances of

success must be severly limi
over most of Europe.

The rounded nose and rather short tail distinguish the water vole,* Arvicola terrestris *from the brown rat. Both are active swimmers but the vole is less shy and will sit in full view to feed on roots and leaves. (size, excluding tail, about 18cm)

Water Vole *Arvicola terrestris*

The reputation of this animal has suffered through constant confusion with the brown rat. In fact the two are dissimilar in all important respects, for although the rat is quite common along urban and rural towpaths, the water vole is more often seen and is more often associated with slow-water habitats.

The rounded nose, small ears and rather short tail of the vole are important features, but these are not always seen if the animal is disturbed. A walk along the canal bank can be punctuated at intervals by the abrupt 'plop' of voles diving from the water edge where they have been resting or feeding. Under these circumstances a good view is virtually impossible, since they will swim underwater until they are among vegetation or until they reach their burrows. These often have entrances both on the bank and below water level, so the voles can reach safety without being seen and can escape in times of danger.

Water voles are territorial, and males in particular usually remain within their own section of water. Male territories are about 130m (430ft), while females take up much shorter ranges and are more inclined to establish new breeding areas.

On heavily-used canals, vole numbers remain quite high, suggesting that the problems of burrow-erosion, water turbulence and pollution are not significant. Certainly voles swim at about the same speed as most canal craft and are able to avoid any obvious danger. They feed on the leaves, stems and roots of waterside plants and are therefore by no means dependent on purely aquatic life.

a zibethicus
of a water vole, the musk rat (known in as musquash) became a disastrous escapee fur-farming days of the 1920s, reproducing in so rapidly that it threatened to undermine the of the River Severn in Shropshire and the Earn in thshire. It reached a peak in 1934, but only three years later had been successfully exterminated by the Ministry of Agriculture.

Coypu *Myocastor coypus*
Once considered to be a severe threat to the dykes of East Anglia because of its tunnelling activities, but at present controlled and confined to east Norfolk and Suffolk.

Coypus are very large rodents, the adults weighing about the same as a fox, and were introduced from South America for their fur. The first escapes were in the 1930s but problems only really developed in the late 1950s. Today coypus are held in check by trapping and the lack of suitable grazing areas close to their burrows. They can swim very well and are often mistaken for beavers, especially when they are seen on river systems away from their normal Fenland range.

Footprints of mink, Mustela vison, *are becoming a common sight along rivers and canals. They are easily mistaken for those of the otter but are only half the size (about 3cm)*

Mink *Mustela vison*
An introduction from North America breeding wild for the first time in 1956 (on the River Teign in Devon). Many people have suggested that mink are partly to blame for the decline of the otter, but since the remains of mink have been found in otter droppings, any interference is likely to be the other way round.

Certainly mink have spread dramatically along river systems and have become pests in some places. Whether they are as damaging as has been suggested by farmers and anglers is open to question. They are very effective killers, as are all the weasel family, and although fish are often taken, moorhens and voles are equally important in their diet.

Mink are mainly nocturnal and are dark brown in colour. As with otters, the best method of discovering their presence is to look for footprints on the water's edge. Interviews with lock-keepers and anglers suggest that canals are as suitable as rivers and that mink will tolerate a high level of disturbance. If numbers increase further it may well prove necessary to control their numbers, but it is difficult to see how this can be achieved — the retraining of otter-hounds to hunt for mink is only likely to succeed in disturbing any otters that are in the neighbourhood. Trapping has already been attempted in some places, but with only limited success.

Slow-moving rivers provide the ideal habitat for the otter Lutra lutra, *but the use of pesticides and the 'clean-up' of river banks have reduced the population to a fraction of what it was thirty years ago*

Otter Lutra lutra

In many ways the otter represents the mystery and challenge of river wildlife. Often researchers and students spend years studying the animal without ever seeing one,

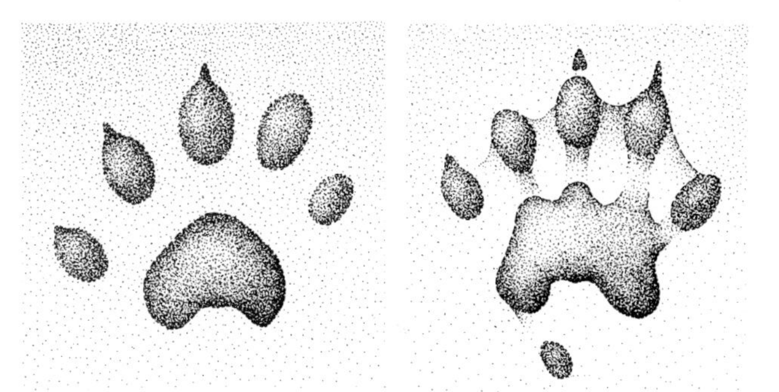

The footprints of an otter in hard and soft mud. It resembles other Mustelids and has five toes, but unlike most of its relatives it has webbing between the toes and this is sometimes visible when the animal has been walking in soft mud. (about 6cm across)

having to base their work on the sightings of tracks, fish remains and, more especially, spraints (droppings) left as territory markers.

Otter numbers declined drastically in the late 1950s, following what may have been a gradual increase in the previous few decades. The reason for the crash was almost certainly the introduction of dieldrin as an agricultural seed-dressing; today it is used considerably less, but it may be that sufficient is finding its way into the river systems to remain an important influence. Otter hunting became illegal in England and Wales in 1977, and although the loss of 22 per year (the reported average) may have been significant during the twenty years prior to this it is unlikely that hunting played a part in the initial decline. Loss of habitat is now considered to be the most important factor preventing recovery.

Otters are essentially creatures of rivers, marshes and (in Scotland) the coast; canals do not figure at all in their distribution though they may be used where linked to rivers or close to more secluded water. Sightings are almost always by accident, for although they take up territories these are quite extensive: males may travel six miles per night in search of food (usually fish or other aquatic animal life).

In many parts of England otters live a nomadic, solitary life in a constant search for suitable habitat, and it is unlikely that numbers will increase until positive steps are taken to create better river conditions.

A dog otter is longer than a fox and twice its weight, so identification is not difficult. Many anglers confuse them with mink, but the latter are only half the size — 60cm (24in) as opposed to 120 (48in) for an average otter.

Common Seal *Phoca vitulina*
Grey Seal *Halichoerus grypus*

Seals are, of course, sea-going mammals, but they are regularly seen in estuaries and tidal rivers. The common seal reaches London every few years and the Nene and the Ouse are often penetrated twenty or thirty miles inland, despite pumping stations. Inland records from the Tyne and other northern rivers may as easily refer to juvenile grey seals which have become disorientated. Seals always create a great deal of excitement, though anglers would find them a considerable nuisance if they appeared in any numbers. One grey seal has been seen to kill thirteen salmon in a single hour!

9 *Waterside Plants*

In some canals and rivers the reedswamp or tall fringing aquatic vegetation described earlier is not present, because of wash or flow, leaving just a marshy zone. Further away from and above the water's edge, the marsh plants cannot compete with grassland species on drier ground. In addition there are muddy and shingle beaches, trees, wet woodland, hedges, walls and bridges to add variety. In towns there are parks where exotic trees line the banks and areas where cultivated plants like horseradish, *Armoracia rusticana,* and garden flowers have colonised the waterside.

MARSHES

Where reedswamp is present, there is often a gradual change, from dense reedswamp to marsh, with mixed intermediate phases. Marshes are best defined as areas where the water level is normally near the surface in summer and which, therefore, are flooded in winter. Several habitual marsh plants have air spaces (aerenchyma) very near the surface to cope with waterlogging. An example is gipsywort, *Lycopus europaeus,* the jagged leaves of which often attract attention before the small, white whorls of flowers. It is common south of a line from the Humber to the Mersey and like other marsh plants including its relative water mint, *Mentha aquatica,* it is

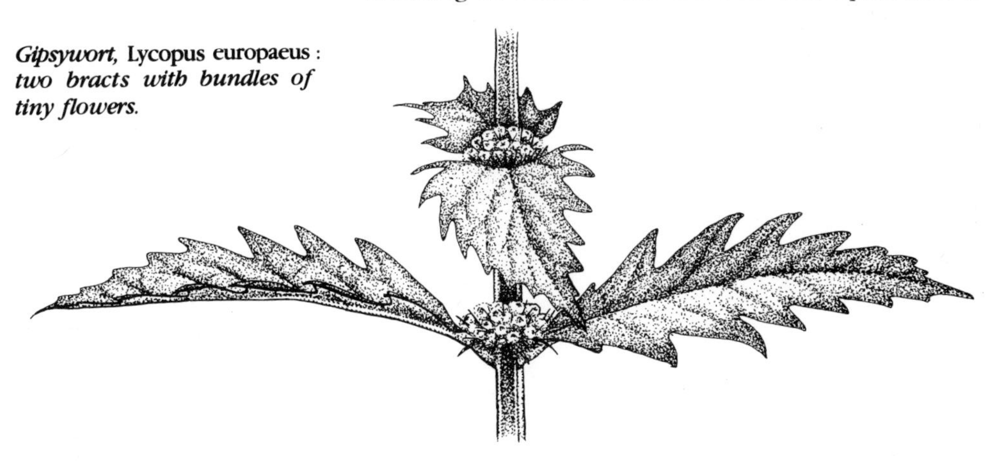

Gipsywort, **Lycopus europaeus** : ***two bracts with bundles of tiny flowers.***

often found out in the swamp zone. Water mint, when crushed or on hot days, gives off the overpowering, aromatic smell which is so characteristic of marshes. The marsh zone can be quite wide in derelict canals, oxbows and backwaters and in places where the railway closely follows the same line as the canal the narrow strip between them often developed into a marsh. Marshes on agricultural land have been drained and many typical plants, like ragged robin, *Lychnis flos-cuculi,* are becoming less common. However, in many canal and riverside marshes it is still possible to see the variety of smaller plants nestling amongst the larger, more showy ones, buzzing with insects on a summer's day.

Skull cap,* Scutellaria galericulata, *on the wall of a lock basin

***Great hairy willow-herb (also known as codlins and cream).* Epilobium hirsutum**

Marshes - Tussocks and Rushes

Many different plants thrive or can tolerate life in a marsh, and some such as common skullcap, *Scutellaria galericulata,* with its bright blue flower tubes grow in reedswamp, marshes and even on the tops of lock gates. Others have very precise requirements of water levels and nutrients and some sedges, for example, the tufted sedge, *Carex elata,* form distinct bands where conditions are just right. Tussock-forming plants are most conspicuous away from flow, wash and trampling, where they can attain their full height; tussock sedge, *Carex paniculata,* can form tussocks 1.3-1.6m (4-5ft) high. Tufted hair-grass, *Deschampsia cespitosa* also occurs near the water's edge but false fox-sedge, *C. otrubae,* is usually higher up the bank. Hop or cyperus sedge *C. pseudocyperus,* a southern species which produces fruits like cones, can also build up tussocks in wet woodland as well as in the open. Rushes are often a sizeable but ignored part of the marsh community and like sedges are often avoided because of the apparent difficulties of identification or their unattractiveness. In fact only six species are widespread and common in this habitat: jointed rush, *Juncus articulatus;* soft rush, *J. effusus;* toad rush, *J. bufonius;* hard rush, *J. inflexus;* and two which prefer acid conditions: conglomerate (compact) rush, *J. conglomeratus* and sharp-flowered rush, *J. acutiflorus.*

Marshes - High Level

One of the commonest of the tall plants is great hairy willow-herb, *Epilobium hirsutum,* or codlins and cream, which can form dense patches of downy plants up to 1.6m (5ft) tall. It thrives in high nutrient levels, either from fertiliser run-off or sewage and can become dominant, crowding out smaller plants. The other willow-herbs found in damp places are smaller. The frothy, creamy flowers of meadowsweet, *Filipendula ulmaria,* often found with codlins and cream have a delicate fragrance and it was used on the floors of mediaeval houses. Another unmistakable tall plant is the marsh thistle, *Cirsium palustre,* which can be over 2m (6ft) tall. The similar welted thistle, *Carduus acanthoides* is much more branched with wings on the lower stems and grows on drier banks. Valerian, *Valeriana officinalis* and wild angelica, *Angelica sylvestris* also have flowers that attract insects. The deep pink buds of valerian develop into a loose head of pink flowers which may be almost white in shade. It may be 1.3m (4ft) tall but its relative, marsh valerian, *V. dioica* is much smaller, has separate male and female plants and grows only in marshes. The native wild angelica is widespread and has smoother stems than the introduced *A. archangelica* which is cultivated to produce the green cake decoration and which grows beside the Thames and in a few places in the Midlands.

The bedstraws, great marsh, *Galium elongatum* which is common in reedswamps, lesser marsh, *G. palustre,* which occurs in marshes and the less common fen bedstraw *G. uliginosum* are all supported by taller vegetation. Another straggler, typical of marshes is the large birdsfoot-trefoil, *Lotus uliginosus,* a bushier version of the common, grassland species, eggs and bacon *L. corniculatus.*

Marshes - Middle Level

One of the most well known marsh plants is kingcup or marsh marigold, *Caltha palustris,* the startling yellow may-blobs, found in or out of shade. Even after flowering the dark, kidney-shaped leaves and the curved fruits are distinctive. The flowers are made of sepals, the outer parts, not petals. Another member of the buttercup family is the upright celery-leaved crowfoot, *Ranunculus sceleratus,* found on damp mud and in marshes generally, but less frequently in the north and west. The widespread yellow loosestrife, *Lysimachia vulgaris,* a member of the primrose family, has a rarer relative, the tufted loosestrife, *L. thyrsiflora,* with small feathery flowers, which occurs in the north and along the Scottish Union Canal. The unrelated purple loosestrife *Lythrum salicaria* has handsome spikes of red-purple flowers and is often found out in the reedswamp zone; like many in this group it is less common in the north.

The figworts, although tall are inconspicuous because the small globular flowers are dingy brown or red. Two grow in damp places, water betony or figwort, *Scrophularia auriculata* which has square winged stems, and figwort, *S. nodosa,* which lacks the wings. A close relative is the pollution-tolerant monkey-flower, *Mimulus guttatus,* a creeping perennial with large showy yellow flowers, spotted with red. The rarer and northern blood-drop emlets, *M. luteus* has bigger blotches. Monkey-flower is found in marshes but also in woods, on weirs, walls, lockgates and even old boats.

The curious flowers of figwort, **Scrophularia** ***spp, which are very attractive to wasps***

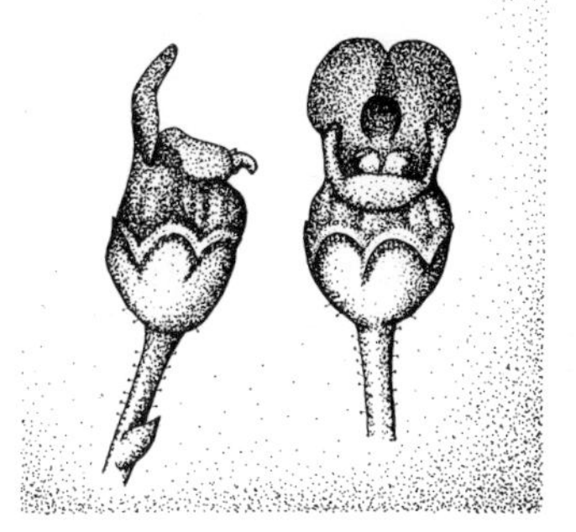

Purple loosestrife, Lythrum salicaria : *one of the most richly coloured plants of the waterside*

The cress family contains several groups of waterside plants, see also pages 59 and 137. The smaller ones may be missed in a casual glance but the flowers of large bittercress, *Cardamine amara* and cuckoo flower or lady's smock *C. pratensis* are easily seen. Large bittercress, which often occurs in alder woods has white flowers with violet stamens, which contrast with the familiar pink flowers, with yellow stamens, of lady's smock, which is more widespread and common.

One group of plants that often receive an unfair share of attention from botanists and conservationists are the orchids. They do have fascinating life histories, not all of which are fully understood and some are very rare, but many other plants are equally deserving of study and conservation. The marsh orchids are a difficult group, not only varying in leaf and flower shape within a species but also crossing with each other, producing fertile hybrids. The fen orchid (or southern marsh orchid) *Dactylorhiza praetermissa* and the northern fen (or marsh) orchid *D. purpurella* only overlap in a narrow band from North Wales across England. The early flowering, marsh orchid, *D. incarnata,* is less common than the other two although it occurs throughout the British Isles. The leaves of pure specimens of all three species are likely to be unspotted, although the northern fen orchid may have round spots. The common spotted orchid, *D. fuchsii,* which has paler flowers and blotched leaves is more widespread and common than the marsh orchids, which prefer wetter conditions. The marsh woundwort, *Stachys palustris,* a member of the nettle family has dense spikes of beetroot coloured flowers flecked with white which bear some resemblance to the marsh orchids.

Marshes - Ground Level

It is very easy to concentrate on the tall, colourful plants but it is always worth looking at the ground for plants and animals. There may be the round leaves of marsh pennywort, *Hydrocotyle vulgaris,* found usually on acid soils or the smaller, paired leaves of bog pimpernel, *Anagallis tenella.* The flowers of pennywort are pinkish green and inconspicuous whereas those of the pimpernel are pale pink and bell-shaped, opening in the sun. The other plants encountered at this level are mainly mosses and liverworts.

The two groups are related and may be found on soil, the bases of trees or on stone structures. Mosses and liverworts are simple plants, without true roots, although special hairs, rhizoids, anchor them; their spores are produced in stalked capsules. Mosses have branching stems with numerous leaves with a mid-rib, but liverworts have lobed leaves without a mid-rib spreading horizontally over the surface. Mosses and liverworts are usually found in wet places where shade or constant supply of water prevents drying out and therefore they also occur commonly under trees. Typical mosses found on the banks on soil or mud include *Ceratodon purpureus, Campylopus pyriformis, Physcomitrum pyriforme* and members of the genus *Bryum,* all of which form cushions or tufts, and *Eurrhynchium* and *Brachythecium* species which are more straggling. Others such as *Amblystegium serpens* and *Cratoneuron filicinum* are found at the base of trees, with *Dicranoweisia cirrata* often found on elder trees.

On mud in shade or on bridges the liverworts form flat cushions. *Lunularia cruciata* has crescent or moon-shaped cups and *Marchantia polymorpha* has large lobes and distinctive parasol-like reproductive structures. Several other species are commonly found but are smaller than *Marchantia.*

OTHER WATERSIDE PLANTS

Several plants, which are not truly marsh species nevertheless choose to live beside canals and rivers and in other damp spots. Two introduced balsams are often found almost in the water; the most spectacular is Himalayan balsam, *Impatiens glandulifera,* with its large mauve-pink flowers, the shape of which give it the common name of policeman's helmet. The seed capsule

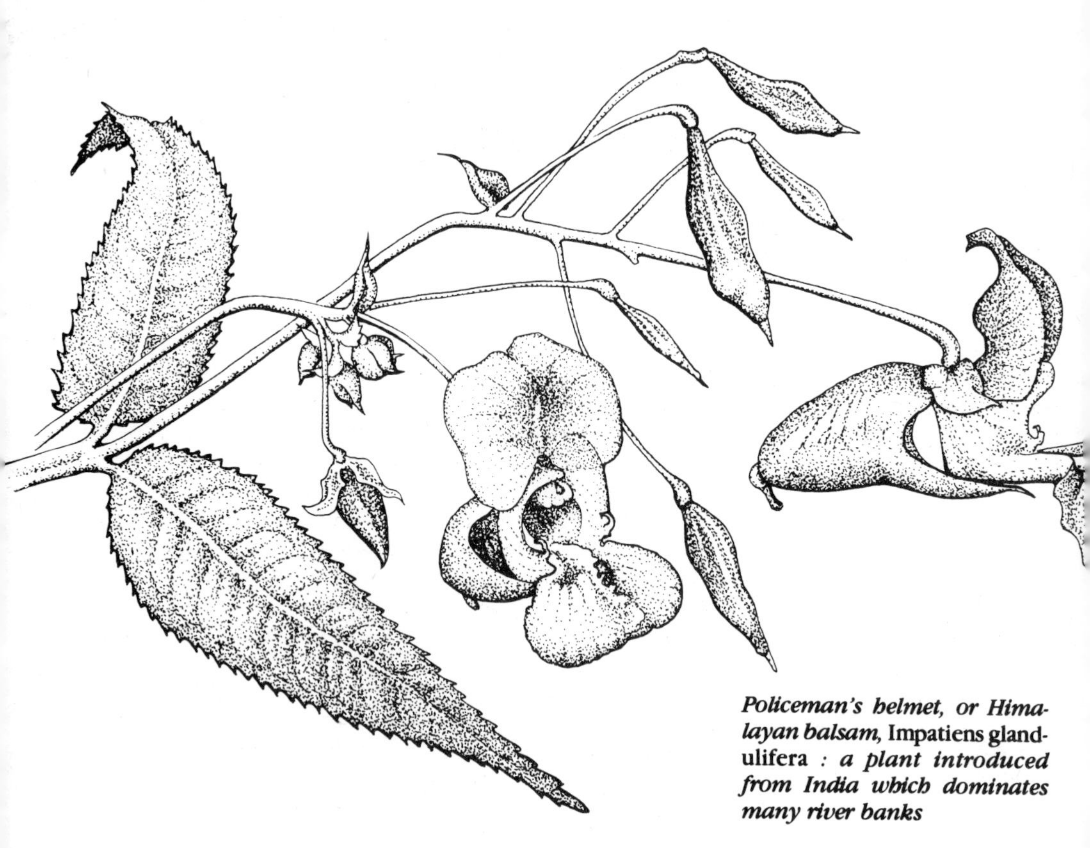

Policeman's helmet, or Himalayan balsam,* Impatiens glandulifera *: a plant introduced from India which dominates many river banks

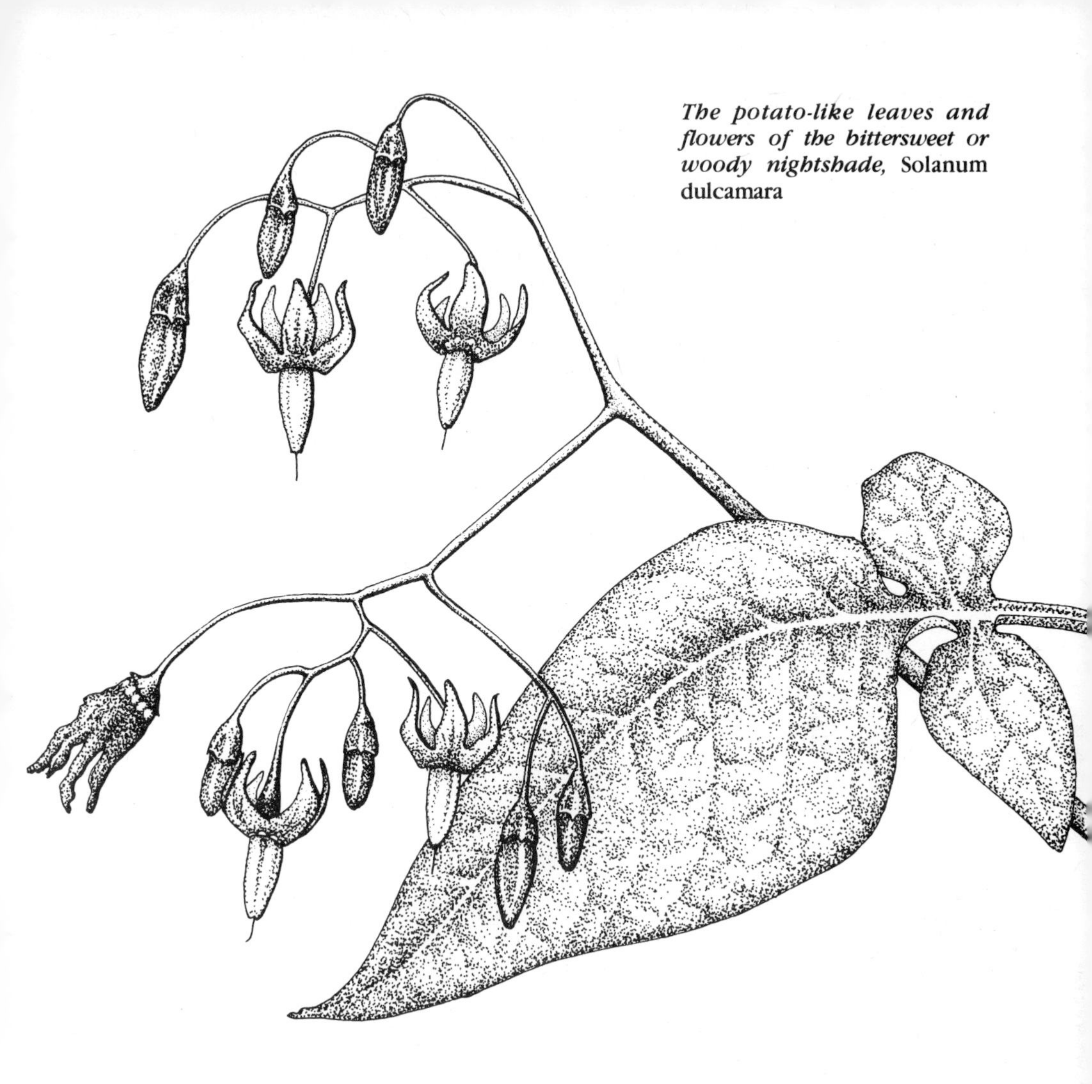

The potato-like leaves and flowers of the bittersweet or woody nightshade, Solanum dulcamara

Orange balsam, Impatiens capensis *: a species introduced from America*

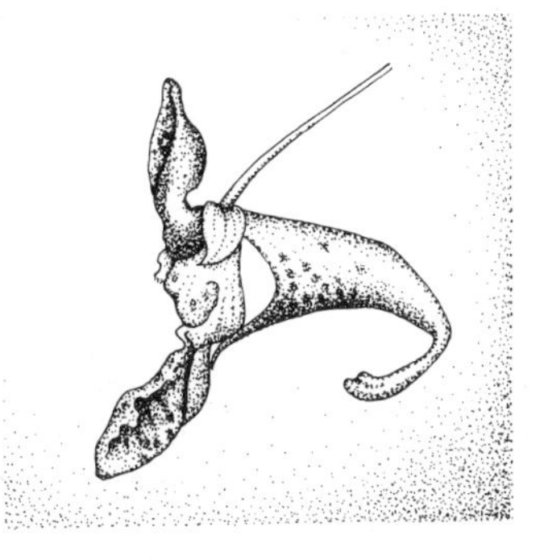

opens explosively in the wind or when touched, shooting the seeds out, causing the rapid spread of the family. Orange balsam or jewel-weed, *I. capensis,* is smaller and has spread into south and central England and Wales. Touch-me-not, *I. noli-tangere* is probably native in the Lake District, but even there it is not very common.

The tangle of tough stems of woody nightshade, *Solanum dulcamara,* is a common sight on banks, often smothering other plants and occasionally blocking small streams by growing out into the water. The purple flowers produce scarlet berries which, like those of the related potato, are poisonous although rarely fatal.

The large rhubarb-like leaves, often 1m (3ft) across, clothing banks puzzle many people. They belong to the butterbur, *Petasites hybridus* which produces its stubby spikes of pink flowers before the leaves come out. The plants are either male or female and the rhizome is

Butterbur, Petasites hybridus *in spring, the flowers appearing before the leaves*

Butterbur, P. hybridus, *in the summer : the flowers have gone and the leaves now resemble rhubarb*

important in spreading the plant and creating the dominant stands, because although males are widespread female plants are rare outside Lancashire, Yorkshire, Cheshire and Derbyshire. Another plant where the leaves may be most noticeable is comfrey, *Symphytum officinale,* the native species with either cream or purple flowers. Although there are other species introduced as fodder crops and for medicinal purposes most of them prefer drier ground and it is the native plants that can come to dominate some river banks. Hemp agrimony, *Eupatorium cannabinum* is another damp-loving plant, common in southern England and Wales, but thinning out as one goes northwards. The tufts of loose heads of pink-purple flowers attract a variety of insects especially butterflies like red admiral and small tortoiseshell. Its specific name, listed in a nature trail has led to a misplaced raid by the drugs squad, but it has no such properties and was considered to have very few herbal uses.

Sweet cicely, *Myrrhis odorata,* a tall, white umbellifer, which smells of aniseed when crushed, also prefers damp soils. It is a northern plant, reaching only as far south as Derbyshire and North Wales and is often found with the beautiful blue meadow cranesbill, *Geranium pratense,* which reaches further south. The other flowering plant included here, the small teasel, *Dipsacus pilosus,* is more like a white scabious than its better known larger relative. It is not very common, occuring in damp woods, often on calcareous rocks but there are several sites by canals.

Waterside Ferns

Some ferns prefer shady damp places although others can grow in the open on river and canal banks. The presence of damp mud is important because ferns have two distinct stages in their life cycle. The prothallus, which resembles a simple liverwort, grows on damp mud and a separate fern plant develops following asexual reproduction. The fern grows and produces sexual spores, often hidden by the scales on the underside of the fronds; these disperse and from them the prothallus germinates. Three ferns are commonly found in damp, often shady places; male fern, *Dryopteris filix-mas,* common buckler fern, *D. dilatata* and lady fern, *Athyrium filix-femina.* All of them form clumps and lady fern is usually found in damper places than the other two, although the three are often found side by side. They differ in the division of the fronds, with male being the simplest and common buckler fern the most complex. The largest British fern, royal fern, *Osmunda regalis,* bears its brown spores on special fronds in the centre of the clump and can be 1.6-2m (5-6ft) high. It was once more common but collectors during the Victorian fern craze and later drainage have reduced its range and it is now only found in the north and west, along rivers and in marshes.

TREES AND WOODS

The main groups of trees and shrubs which occur on the banks of rivers and canals and in marshy places are the birch family, which includes alder, *Alnus glutinosa,* and the willow family which contains willows, sallows and poplars. Both families bear separate male and female flowers; in birches male and female catkins occur on the same tree whereas in the willow family they are on different trees. Trees, for example oak, *Quercus robur* and ash, *Fraxinus excelsior,* which prefer drier habitats also grow on river and canal banks, as isolated trees or in woods but are not characteristic of the habitat.

Reptiles are not usually associated with water, but the grass snake, **Natrix natrix,** ***swims well and preys extensively on frogs*** ▷

Alder, **Alnus glutinosa** ***: male catkins forming in the late summer***

Alder and Birches

Alder is probably the commonest waterside tree, lining mile upon mile of bank and also forming the distinctive wet woodland, carr, either alone or with a mixture of hairy birch, *Betula pubescens,* willows or sallows. It can be a majestic tree, up to 20m (65ft) tall but it is often cut down, to prevent it from becoming top-heavy and damaging the banks, and then allowed to regrow to encourage the roots to bind the bank. The leaf and catkin buds are purple and the green, female cones turn brown and woody as the nutlets enclosing the seeds develop. The nutlets have air pockets to help them float once they fall, and the old empty cones stay on the tree alongside the new green ones.

Tussock sedge is a common constituent of the ground flora of alder carr; other smaller sedges likely to be found are the ubiquitous common sedge, *Carex nigra,* remote sedge, *C. remota,* and common wood sedge, *C. sylvatica* or its lowland, wet-loving relative *C. strigosa,* which is rarer. The lesser water forget-me-not, *Myosotis laxa* ssp *caespitosa,* is more common in carr than in the open, as is golden saxifrage, *Chrysosplenium oppositifolium,* which grows on shady stream banks and spring lines, forming bright yellow cushions when the petal-less flowers appear early in the year. Its rarer relative, *C. alternifolium,* prefers basic soil, although they occasionally grow together. Other

▷ ***Fish are only part of the diet of the mink,*** **Mustela vison,** ***which is now one of the most widespread riverside predators***

marsh plants which tolerate or prefer shady places like the yellow pimpernel, *Lysimachia nemorum,* and the introduced pink purslane, *Montia sibirica,* are also found in carr.

Silver birch, *B. pendula,* is a very familiar tree but it is the hairy birch which grows better on wet soils. Unlike alder, birches have relatively smooth, acid bark, which discourages the growth of epiphytic mosses and liverworts, although birch woods and the trees themselves support a range of fungi. The large bracket fungus, *Piptoporus betulinus* is white or grey at first, darkening later and is specific to birch. Another fungus, *Taphrina betulina,* causes witches' broom galls. It appears that bud production is encouraged, but growth is inhibited by the fungus until it becomes less active, at which point the shoots develop to form the 'broom' which can contain over one hundred twigs.

Willows and Sallows

Willows, like birches, hybridise and cause confusion in their identification; it did not help that the separate male and female trees were originally classified as separate species, before their relationships were worked out. Only nine species are commonly found by water, either as individuals or in woods. The shrubby, broad-leaved great sallow (or goat willow) *Salix caprea,* and common sallow, *S. cinerea,* produce their flowers before the leaves and the male catkins are the pussy willow collected for Palm Sunday. Three other wetland species, eared sallow, *S. aurita,* osier, *S. viminalis,* and purple osier, *S. purpurea,* also do this but they have poorer catkins or flower too late to attract attention and in the other species the leaves appear first.

Only two willows make sizeable trees, up to 25m (80ft) high and both have narrow, lanceolate leaves. Crack willow, *S. fragilis* is more common than white willow, *S. alba,* although both are widespread except in the north and west. They are both pollarded, but when they grow naturally the crack willow has more spreading branches, which break easily in wind. It has striking red or pink roots which are often visible, whereas the white willow has white roots as well as the silvery white leaves which give it the name. A sub-species of white willow is used for making cricket bats, ironically from female trees which produce the better timber. The only other willow with similar leaves is the osier, which is much smaller, rarely 5m (16ft) tall, and is commonly cultivated and managed in beds, because its pliable, straight wood is used for basket-making.

The widespread sallows have broader leaves and except for eared sallow, which only grows 1-2m (3-6ft) high, are usually about 4.6m (15ft) tall although they can reach 9m (30ft). Eared sallow is the easiest to recognise with the

◁ *The monkey flower,* Mimulus guttatus, *is an attractive American plant, first recorded in 1830 and now common along riverbanks throughout Britain.*

◁ Rosa villosa - *a northern rose of exceptional beauty. The leaves are slightly downy and the thorns straight and very sharp*

'ears' on the leaf stalks and it is most common in north Scotland, although widespread throughout Britain. Common sallow is more frequent in damper spots than great sallow, which is found in drier sites than most other willows. Under the bark of older twigs of common and eared sallows there are ridges which are absent in great sallow.

The northern bay-leaved willow, *S. pentandra,* has glossy, dark green leaves and is often the latest to flower. The almond willow, *S. triandra,* which is scarce west of a line from the Humber to the Severn Estuary, grows taller than the more common purple osier although both were commonly planted for basket-making.

The weeping willow, *S. babylonica,* is an introduced species, originating in China and south-east Asia, not Babylon, but a cultivated weeping variety of white willow is also planted in private and public gardens beside canals and rivers.

Poplars

Three species of poplars are thought to be native. The grey poplar, *Populus canescens,* is probably only indigenous in south-eastern England. It grows in damp or wet woods and like the black poplar, *P. nigra* can grow to more than 30m (100ft). The native black poplar only suckers rarely, has deep cracks and bosses on the bark and its branches tend to curve downwards. It is restricted to eastern and central England, where it grows by ditches and streams and also in woods; in some springs the red anthers of the catkins colour the male trees. Identification is complicated by the range of widely planted hybrids, although they have upward pointed branches, which is carried to the extreme in the Lombardy poplar, which lines stretches of large rivers, such as the Thames.

Aspen, *P. tremula,* has rounder leaves than the other poplars, especially on the numerous suckers which enable it to become dominant in places. The lateral flattening of the leaf stalk which is characteristic of poplars is exaggerated in aspen, allowing free movement, the 'trembling' which gives it the latin name. It will grow on drier land but also on wet acid or neutral soils and is commonest in the south-east although generally distributed.

In damp areas by water courses there are many small poplar plantations, put in to dry the ground, or as part of a campaign to supply the match industry with timber.

One shrub not belonging to the characteristic tree families is alder buckthorn, *Frangula alnus* which favours damp woods in southern England. It is not particularly attractive and is often overlooked, but it is important as one of only two food plants for the brimstone butterfly caterpillars.

Pollarded willows on the Trent and Mersey Canal, Stenson, Derbyshire

BARE GROUND

There are other drier habitats and although they are common elsewhere they are worth mentioning in connection with rivers and canals because they add not only to the attraction of a walk beside a river or canal but also to the wildlife communities. Bare ground is common by rivers where floods or normal erosion create shores and cliffs and on canals where there are piles of dredgings. Some of the colonising plants are common elsewhere, but others like damp mud and do well by water. Coltsfoot, *Tussilago farfara,* is one of the first colonisers of bare earth and its sulphur-yellow flowers appear long before the heart-shaped leaves, like its relative, butterbur.

Silverweed, *Potentilla anserina,* with its distinctive silvery leaves and yellow flowers often covers drying mud exposed as the water level drops in spring. The same conditions are ideal for the germination of marsh yellowcress, *Rorippa palustris,* the less common creeping yellowcress, *R. sylvestris* and the yellow winter cresses, especially the common *Barbarea vulgaris.* A flowering plant which people mistake for maidenhair fern is the lesser meadow rue, *Thalictrum minus* which thrives on gravelly patches by the water's edge. Another habitat which can be important is the old sites of fires, where branches and cut grass have been burnt. The most noticeable plants in the early stages after the fires are the mosses, especially *Funaria hygrometrica,* a pale green cushion moss which produces orange-yellow stalks with spore capsules. Later on rosebay willowherb, *Chamaenerion angustifolium,* colonises the site.

GRASSLAND HABITATS

Grazing, cultivation or the use of herbicides often reduces the wildlife interest of the banks but on the towpath side of canals and along certain stretches of rivers there is rough grassland with a variety of plants. Some cutting or grazing is necessary to stop the taller species overwhelming the smaller ones; the grassland that has developed depends not only on the previous and present management but also on the geology of the area. Thus chalk or limestone meadow plants which are now rare on the adjoining farmland are often found beside canals and rivers. Some of the plants of wet meadows such as fritillary or snakeshead, *Fritillaria meleagris,* have declined rapidly in recent years. Fritillary was found in twenty-seven counties in one hundred and sixteen 10km squares before 1930; by 1977 it was only found in quantity in nine counties in 15 squares in southern England and the Midlands. Most of the losses are the direct result of drainage, although picking, grazing during the flowering period and bulb-digging take their toll. Another species affected in the same way is the summer snowflake, *Leucojum aestivum,* which occurs in wet meadows and willow thickets along the Thames Valley.

HEDGES

Occasionally, there are reminders of the woodland origins of our landscape, especially on canal banks, where hedges provide shade and shelter for primroses, *Primula vulgaris,* and violets, *Viola* species. The hedge itself is very important as a habitat, providing a variety of food, shelter and nesting sites for birds, from the ground through the hedge shrubs to the hedgerow trees. Insects feed on the

A hawthorn hedge, recently cut and layed, and a flower-rich towpath : the Grand Union Canal near Weedon in Northamptonshire

hedge-bottom plants and on the twigs, flowers and fruits of the shrubs. Small mammals use the corridor under the hedge and eat berries and nuts. A hedge can also be important for the wildlife on the bank, sheltering it from wind and rain and the plants and animals on one side may be different from those on the other side where conditions may be colder and wetter.

Hedges are found by rivers but they are more characteristic of canals, which were cut through existing fields; hedges were needed to stop cattle and sheep straying along the towpath. The main period of hedge planting under Parliamentary Enclosure Acts lasted from 1760 to 1820, overlapping the canal building era. Therefore, nurserymen were already producing hawthorn (quickthorn), *Crataegus monogyna,* blackthorn, *Prunus spinosa,* and timber trees such as oak, *Quercus robur,* ash,*Fraxinus excelsior,* and elms, *Ulmus* species, which were the main species planted at this time.

Four shrubs which are less commonly found in hedges are guelder rose, *Viburnum opulus,* dogwood, *Cornus sanguinea,* spindle, *Euonymus europaeus,* and the native field maple, *Acer campestre;* all are characteristic of rich or basic soils. Guelder rose is related to elder, not roses, and the attractive flat white flower head has infertile flowers surrounding smaller fertile ones which develop into scarlet berries matching the vivid autumnal colours of the leaves. The red twigs of dogwood and the green ones of spindle stand out in winter and the hardness of both woods led to their use for skewers. Spindle is most striking in autumn when the lobed pink berries open to reveal the orange-coated seeds, and nature gets away with this colour clash. It has been eradicated from many hedges because it harbours the eggs of the pest, black bean aphid.

A conspicuous group of hedgerow plants are the climbers and clamberers, which use other plants to raise themselves off the ground. The only ones that twine in an anti-clockwise direction are great bindweed, *Calystegia sepium* with the large white trumpets, and field bindweed, *Convolvulus arvensis.* Clockwise twiners include honeysuckle, *Lonicera periclymenum,* hop, *Humulus lupulus,* and black bryony, *Tamus communis.* The oval leaves of honeysuckle are visible for most of the year, often appearing in December and not dying until late autumn when the tight cluster of red berries has developed. The tough twining stem can severely contort the supporting branches and attractive walking sticks are sometimes made from hazel affected in this way. In hop, black bryony and white bryony, *Bryonia dioica,* which are all totally unrelated, the male and female flowers are on different plants, a feature shared with holly. The bryonies are the only native representatives of two exotic families, yams (black bryony) and melons (white bryony). At first sight the chains of poisonous red berries of the bryonies are similar, but white bryony climbs by using tendrils, which grow from the stem, not the end of the leaves as in some vetches. White bryony is a southern and eastern plant whereas black bryony, which needs moisture for germination, extends into the west including Wales.

Traveller's joy or old man's beard, *Clematis vitalba,* best known for its feathery fruits, climbs by twisting its leaf stalks around other plants; the leaves are pinnate, unlike all other climbers which have simple leaves. It is another southern species, found mainly on chalk and limestone south of the River Humber. Ivy, *Hedera helix,* has found another way of climbing, using 'rootlets' which do not parasitise the tree; otherwise ivy would not survive when climbing walls. The flowers, which appear in September are a rich source of pollen and nectar and provide a welcome late food supply for insects and then for birds when the berries ripen later. It is the only truly evergreen climber and provides shelter for birds, insects and small mammals through the winter. The uninviting, tough leaves of both ivy and holly are the alternate food of the caterpillar of the holly blue butterfly.

The main scramblers are roses and brambles which, as everyone knows, are effective barriers. The dog rose, *Rosa canina,* is the most widespread although the field rose, *Rosa arvensis,* and sweet briar, *Rosa rubiginosa,* both occur in the south, to be replaced by the downy rose, *Rosa villosa,* in the north. The yellow-red spiky growth on some roses is a gall caused by the larva of the small wasp, *Diplolepis rosae,* but its predators and some parasites also shelter in the 'briar ball'. Over three hundred brambles have been differentiated, according to leaf shape, the numbers and types of glands and prickles, the flowers and fruits. Some are good to eat, others are bitter or produce poor fruits and in any picking expedition it is worthwhile tasting the fruit and being selective.

Finally, there are other minor habitats, which occur elsewhere but which can be an interesting feature of a waterside walk. These are the stone and brickwork of bridges, locksides, tunnels and walls and the lock gates. Although plants from other habitats can grow there, those that are typical of dry walls have rosettes of leaves, are

A derelict lock on the Grand Union, superceded by a double basin to the left. The old lock chamber has been colonised by a fine assortment of plants - particularly the great water dock, Rumex hydrolapathum

cushion-shaped or are hairy to reduce the loss of water. Striking cushions of yellow biting stonecrop, *Sedum acre,* the trailing, purple flowered, ivy-leaved toadflax, *Cymbalaria muralis,* the western wall pennywort, *Umbilicus rupestris,* ivy, and the more local pellitory-of-the-wall, *Parietaria judaica,* are all characteristic of walls, but small cranesbills, chickweeds and groundsel, *Senecio vulgaris,* manage to cope with the arid conditions on walls. The type of stone may be important, stonecrop preferring basic rocks and wall pennywort more acid ones. Even the basic mortar may have different plants to the rest of the wall. Wall ferns, such as wall-rue, *Asplenium ruta-muraria,* black spleenwort, *A. adiantum-nigrum,* rusty-back fern, *Ceterach officinarum* and hart's tongue fern *Phyllitis scolopendrium* all prefer limestone but may use the mortar. They are all less common in the drier eastern counties. Most mosses found on walls, such as *Tortula muralis, T. subulata, Homalothecium sericeum, Bryum capillare* and *Barbula convoluta,* form cushions to conserve water, and like ferns often occur most frequently on limestone. Several lichens are also obvious on dry walls. They are a peculiar combination of fungus and alga, first recognised by Beatrix Potter and there is much still to be understood about the way the two work together and how they reproduce. It is clear that they are slow growing and a small ten-pence piece sized plant represents several years' growth. The most striking plants on limestone walls are those like the orange-yellow *Caloplaca heppiana* and *Xanthoria* species but small white or grey patches are also lichens.

In tunnel mouths, lock sides and other wet places, liverworts, mosses and sometimes ferns can survive although in tunnels they are restricted to within a short distance of the mouth and below air vents where the light is sufficient. The liverworts and mosses include most of the species already mentioned in the marshy areas as well as willow moss, *Fontinalis antipyretica* and others which, on lock sides, can survive inundation, wear and tear and pollution.

The banks contain much more than just marsh plants and as a meeting place of wet and dry, natural and man-made habitats offer many contrasts and interests.

10 *Waterside Invertebrates*

Although a great many invertebrate animals are restricted to damp conditions very few other than insects are associated with waterside habitats. This may seem a contradiction, especially since many of them evolved from aquatic ancestors and only set foot on land in comparatively recent times. Woodlice are good examples of this; they are crustaceans with no terrestrial relatives, but a whole host in the sea and in freshwater — everything from crabs to water-lice. The need for a humid atmosphere arises from an inability to withstand desiccation, their body-wall being permeable and therefore vulnerable. They live under stones or bark where a damp microclimate prevails, and expose themselves to the atmosphere only at night. Even vegetation growing out of water can prove very dry, particularly in windy conditions, and the alternative of

The Cromford Canal near Ambergate in Derbyshire : an excellent example of a disused canal with rich insect habitats on the towpath side

The banks of the River Avon, above Stratford, have a varied flora and provide good breeding sites for invertebrate animals

total immersion is just as dangerous. Molluscs and insects have overcome the problem and avoided the evolutionary trap, the former by constructing calcium-based shells into which they can withdraw, and the latter by utilising a waxy covering or cuticle over their exo-skeleton. Both are prolific in wetland sites, particularly insects which have managed to dominate the majority of biotypes, wet and dry.

Most inhabitants of river and canal banks associate themselves either with certain plants or zones of plants, and for this reason it is more convenient to consider them by their preferred zones rather than by their taxonomic groups. There are many overlaps of course, particularly with such mobile animals as insects which may spend much of their adult life on nectar-rich flowers some distance from their larval foodplant or breeding site.

The four zones are identified as follows:

1. *Reed layer:* emergent vegetation, including rushes and sedges.
2. *Moss layer:* very short ground-plants, providing a moist and permanent micro-habitat close to the water's edge.
3. *Herb layer:* annual plants growing to a height of a metre or more, often found as a band close to the water's edge but above the water level.
4. *The Shrub layer:* perennial plants up to five metres tall.

Along canals, these zones are often continuous, while along rivers they may be fragmented but still identifiable.

1 Reed Layer

Apart from dragonflies and the other semi-aquatic insects already considered in chapter 7, the most obvious invertebrates of the emergent zone are molluscs. Amber snails, the Succineidae, are very common, *Succinea putris* being a particularly noticeable species, usually pale brown in colour and about 15mm, on the leaf blades of reeds or sedges. There are three other species of amber snails and their identification is virtually impossible without dissection. *Oxyloma pfeifferi* is the only other one likely to be encountered in any numbers however, typically a little smaller than *Succinea* and often almost black in colour, found again on any damp vegetation both above and below the surface. One slug is associated with this type of habitat, a small brown species, *Deroceras laeve,* called the marsh slug and usually abundant among dense rotting stems where it seems quite happy to be under water for long periods.

It is the regularity and the duration of immersion that causes problems for most insects attempting to make use of the same habitat. If they remain exposed on the leaves they will be picked off by birds, and if they withdraw down to the base of the reeds, which is the usual strategy, they will become too wet and open to fungal attack. Many 'fen' insects, like the large copper butterfly, *Lycaena dispar,* and the water ermine moth, *Spilosoma urticae,* have larvae which display some resistance to flooding and may remain exposed on leaves where they rely on either camouflage or unpalatability (ie they may taste unpleasant or be very hairy). Riverine habitats are very different from fens, however, and the amount of cover is considerably less, so exposure has to be avoided. The dormant winter stage usually has some built-in insurance against mould or the attention of predators, and may be spent in any stage of the life-cycle. Fen species have adapted themselves to withstand quite long periods of flooding, and while riverine species do not have to cope with such extreme periods of immersion, they have to contend with the danger of being swept away in flash-floods and winter spates.

Not surprisingly, there is a considerable overlap between species of the reed zone of rivers and canals and those of fens, but while most riverine species are found in the classic fen sites of East Anglia, like Wood Walton or Wicken, the reverse is not often the case. This is why the drainage of fenland has caused the extinction of so many insects. The short-winged cone-head, *Conocephalus dorsalis,* a small but very elegant bush-cricket, is one of the few examples of a fen-type species which has managed to survive in marginal habitats and is found all around the south and east coasts where lowland rivers and marshes have provided alternative sites. More specialised fen species, like the reed tussock moth, *Laelia coenosa,* have not been so adaptable and have vanished over the past 150 years.

Moths of the wainscot family have evolved a very successful stategy that has enabled them to avoid many of the problems outlined above. Their larvae feed on marshland plants, and several of the larger species, like the bulrush wainscot, *Nonagria typhae* and the large wainscot *Rhizedra lutosa,* lay their eggs in the stems and roots of emergent plants, the larvae developing inside tunnels excavated in the pith. Webb's wainscot, *Archanara sparganii,* takes this process one step further, starting its life in the stem of a bur-reed but transferring to *Phragmites* reed when it becomes too fat and needs a stem of greater diameter. All the wainscot moths are night flying, emerging in the late summer or autumn, and while several are confined to the south and east, many occur wherever the habitat and foodplant are to be found. Thus the large wainscot, which tunnels into reed stems, is common well into Scotland. Like most of the true 'fen' species, the wainscots are pale powdery brown in colour and may look almost white when caught in a beam of light. A riverside on an autumn evening is therefore a particularly productive habitat in which to use a torch, but not a very good place for trying to catch anything beyond arm's reach.

Several flies are associated with reeds, but the majority of these, such as the tiny Anthomyzids, are rarely noticed. The family Chloropidae contains a particularly interesting species, however, called *Lipara lucens,* which causes large cigar-shaped galls on *Phragmites* reed stems. The *Lipara* larva develops inside this gall, feeding in comparative safety in much the same way as the wainscot moths. By providing such an ideal breeding site, the *Lipara* larva plays host to several other flies such as *Cryptoneura flavitarsis,* which does not harm the original occupier but uses the surplus of food. Once the *Lipara* and its guests (called 'inquilines') have emerged, leaving a tunnelled out chamber with a hard outer casing of leaf bracts, the residence becomes attractive to a whole range of other insects searching for safe winter quarters. Chief among these is the bee *Hylaeus pectoralis* which fills the chamber with a line of cells, each provisioned with a 'pollen loaf' and containing a single egg. These hatch and develop in the autumn, the adults emerging the following summer when they visit flowers of the herb zone for the necessary nectar and pollen. A wasp, *Passaloecus corniger,* uses the old *Lipara* galls too, provisioning them with aphids rather than pollen, and is difficult to distinguish on the wing from *Hylaeus;* both are small and black.

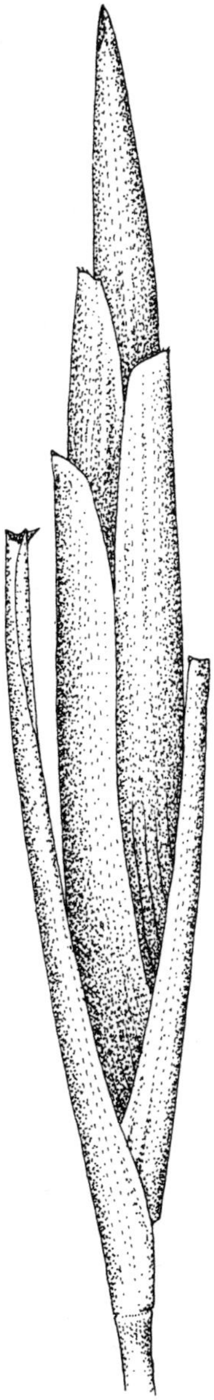

With safe breeding sites at such an obvious premium, it is no surprise to find several insects and spiders using the flower heads and seeds of emergent plants. One particular

***The 'cigar-gall' made by the small fly* Lipara lucens *is sturdy and remains on the stem long after the original occupant has left. The hollowed-out gall is then used by other insects : particularly the wasp* Passaloecus corniger *and the bee* Hylaeus pectoralis**

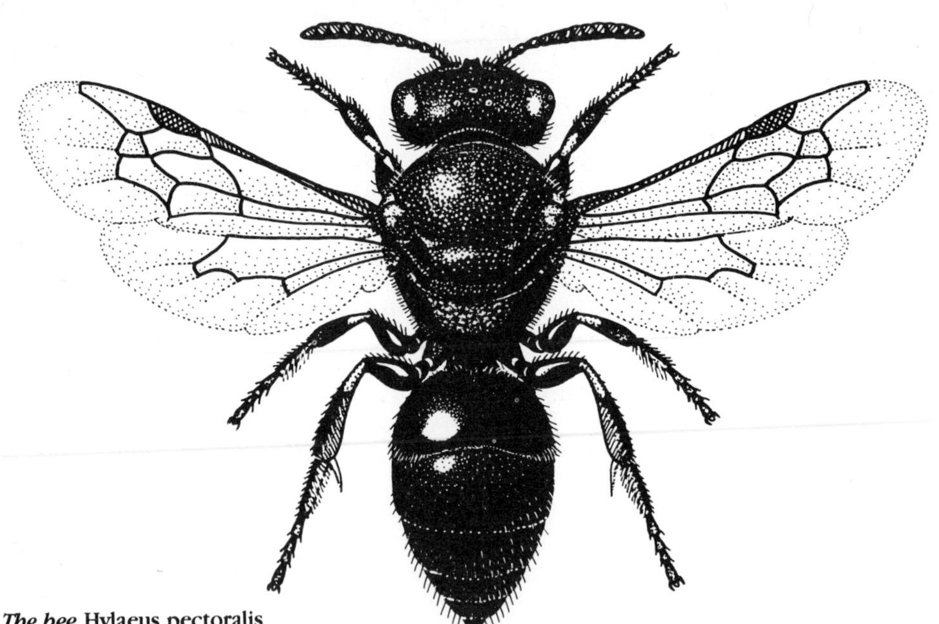

The bee Hylaeus pectoralis
(wingspan about 10mm)

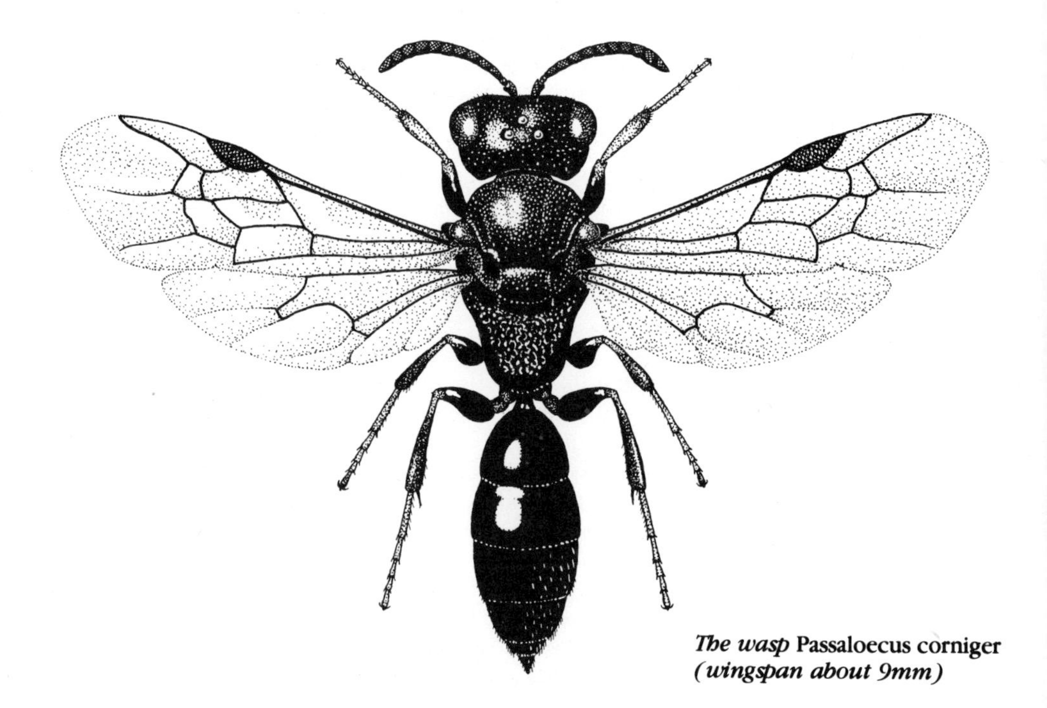

The wasp Passaloecus corniger
(wingspan about 9mm)

specialist worthy of mention is the bug *Chilacis typhae* which spends most of its life in reedmace ('bulrush') heads and is never found far from this secure and abundant supply of food. Most other invertebrates use these sites only temporarily, however, and have arrived from less problematic habitats. The reed zone for much of the year remains a hazardous and temporary resource.

2 Moss Layer

Bare stones and tree stumps close to the water are almost always colonised by bryophytes (mosses and liverworts), which then accumulate a matted layer of wet humus and are an ideal habitat for small insects. Sifting through one of the *Cinclidontus* mosses (*fontinaloides* on the actual water edge and *mucronatus* on alder or willow stumps close by) always produces a confusing assortment of jumping or wriggling creepy-crawlies. Springtails (Collembola), which are tiny purple/grey creatures, are particularly abundant, but the moss is also a favourite hunting ground for rove beetles, (Staphilinidae). These are very small, thin beetles, usually shiny black, which prey on springtails and the young larvae of flies. The majority of insects in the moss layer are only there in their immature stages, however, and spend much of their adult life on higher vegetation. Thus they find food and shelter during the early period of their lives, but disperse to meet mates and lay eggs in new bryophyte clumps. The largest of the British lacewings, *Osmylus fulvicephalus* is a fine example of this. The adult has an attractive brown mottling on its latticed wings and may measure up to 5cm (2in) across, but it is rarely seen unless disturbed from nearby bushes. It flies during the early summer and is one of the small but diverse group of insects in which the female takes on the positive role and seeks out the male. The larva is one of the largest predators found in the moss zone, equipped with a scimitar-shaped pair of mandibles with which it hunts springtails, which are the basic unit in the scale of predation for the whole habitat, but when fully grown it transfers its attention to the larvae of semi-aquatic flies. Old canals, with moss-covered ledges and lock structures, are always rich in these, the mud and ooze providing ideal conditions.

Another example of an insect breeding in the moss-zone, but being more readily observed elsewhere is the moth, *Micropterix calthella*. The genus *Micropterix* contains only six species, all very small but of great interest because they are extremely primitive and a link-group between moths and caddisflies. Instead of the curled drinking-straw system employed by all other moths, the *Micropterix* group have mouthparts containing mandibles, so instead of sipping nectar they are obliged to chew pollen. *Micropterix calthella* measures about 8mm across the wings and is therefore very small, but its habit of congregating in large numbers in buttercup or marsh marigold flowers makes it quite easy to spot. The wings are dark green and are held around the body, and there is an orange tuft of hair on the head which is very distinctive. The main congregation is during the first few days of June, and thousands of insects are involved: presumably they mate on the flowers and the females then return to the moss layer where they lay their eggs. The whole life cycle of all the *Micropterix* moths is obscure and several species still have undescribed larvae. *M. calthella* feeds on moss and liverwort and is common throughout Britain.

Of the non-insect invertebrates, several millipedes and centipedes are to be found under stones and bark, often among moss but by no means exclusively so. The same is true of woodlice, false scorpions and harvest-spiders that make up the rest of the land invertebrates. Of the true spiders, however, one species should be mentioned since it is the largest British arachnid and is quite well known. This is the raft spider, *Dolomedes fimbriatus* (another species, *plantarius* also occurs but it is extremely rare), which inhabits the edge of ponds and streams and is known from several canal sites. It is really an animal of tangled marginal undergrowth, but is known to submerge itself when alarmed and is perhaps best considered as a species of the moss layer, close to the water's edge. It is by no means widespread and has disappeared from many old established habitats in the extreme south, but where it occurs it is often numerous and remains one of the most impressive of our riparian invertebrates.

Full-grown caterpillars of the drinker moth, **Philudoria potatoria.** *(about 70mm)*

3 *Herb Layer*

The lush waterside grasses provide a staple food for many of the larger moths, for example the larva of the drinker moth, *Philudoria potatoria,* which always provides a lot of excitement when found. It is big and hairy and sits high on grass stems to browse slowly on the leaves. Its defence of hairs is proof against most birds except the cuckoo, which seems oblivious to irritation, and small boys who always like hairy caterpillars as pets. Its name is derived from the unusual habit of its caterpillar of noisily drinking beads of dew or rainwater. The most abundant moths of the grasses are again the wainscots, however, (for example the smoky wainscot, *Mythimna impura)* but the larvae are free-feeding and only come out at night. Like the drinker and most of the other grass-feeders, the wainscot moths overwinter in the larval stage and are able therefore to make the most of the early spring growth.

Apart from the grasses, the rich flowering herbs also attract many other moths — for example the butterbur, *Hydraecia petasitis,* which is a northern speciality and burrows in the roots and lower stem of the foodplant. A

One of the most attractive day-flying moths is the scarlet tiger, Callimorpha dominula. *Its caterpillar is equally distinctive and sits in full view on comfrey leaves. (35mm)*

better known species, the elephant hawk moth, *Deilephila elpenor,* is far more widespread and easily found. The larva feeds on rosebay willow-herb, *Chamaenerion angustifolium,* but has recently taken to orange balsam, *Impatiens capensis,* and has thereby extended its range to the canal-edge as well as the bank. The adult is very beautiful, with heavily angled wings and a plumage of pink and olive-green. It is strictly nocturnal, however, and it is the very large caterpillar that is most often seen. This could not be described as attractive in the normal sense, a fully-grown specimen being velvety-brown and hairless, rather fat with a tiny spiked 'tail' at one end and false eye-spots at the other. Its practice of extending its head and waving its front end around like an elephant's trunk is responsible for the otherwise incomprehensible English name. Fully-fed during September, it has a habit of climbing high onto the foodplant to sunbathe during the late afternoon and is then encountered by walkers who are either revolted or intrigued. The transformation from a fat brown 'grub' to an exquisite winged machine is nowhere so well exemplified as in this insect.

Along the water-meadow margins and riversides of the south and west may be found a moth which is beautiful in both stages. This is the scarlet tiger, *Callimorpha dominula,* which is local in distribution but is diurnal both as a larva and an adult, so where it is found it is usually noticed. The larva is black with spangled warts and tufts of shiny black hair, and a series of yellow and white markings along its back and sides. It sits quite openly on the foliage of comfrey *Symphytum officinale* and is never very far from the water's edge. The adult moth, which flies in June, is even more impressive, having deep green-black forewings marked with cream spots, and bright scarlet hind wings. Its colours and habit of flying in the sunshine often result in its being mistaken for a butterfly, and it is certainly one of the most eye-catching of riparian insects.

Waterside butterflies are virtually unknown in this country, and the species met with by rivers and canals are there because of an associated habitat — usually a hedgerow. Along canal towpaths the wall brown, *Lasiommata megera,* is characteristic, while the small heath, *Coenonympha pamphilus,* meadow brown, *Maniola jurtina* and the small copper *Lycaena phlaeas* are also very common. The latter species is associated with docks (*Rumex* species) while the others all lay their eggs among grass.

Two species of butterflies make use of lady's smock, *Cardamine pratensis* as a foodplant — an alternative to garlic mustard, *Alliaria petiolata,* and therefore an overspill from the hedgerow to the bank. These are the orange tip, *Anthocharis cardamines,* and the green-veined white, *Pieris napi,* both of which may be mistaken for 'cabbage whites', though the male orange tip is much more delicate and has the marking suggested by its name.

The abundance of butterflies along riverbanks is often a product of the absence of herbicides, rather than a particular affinity with water, and the same principle applies to many other groups of insects. Most orders have specialised families or individuals, however, and these may be confined by a foodplant or by a particular predator/prey relationship. One of the less attractive is that between the fly *Lucillia bufonivora* and the common toad. The fly, a species of greenbottle, seeks out a toad during the day — presumably by scent since toads lie well hidden until dark — and lays a batch of eggs on its back. The maggots hatch quickly and make their way to the eyes, thence to the nostrils, and begin to eat away the membranes and tissue inside the toad's head. In two or three days the toad dies and the maggots finish off the remains in less than a week. The whole cycle lasts no more than a fortnight. This is certainly one of the most horrific stories of life on the waterway, particularly since the affected toad cannot breathe normally and is obviously in great distress as it blunders about trying to find relief.

A much less insidious relationship is exhibited by the large but fascinating group known as sawflies. These are not really flies at all, being primitive Hymenoptera and therefore closely related to the wasps. They have rather thin exo-skeletons and are particularly common in damp places. Many are associated with trees (see *Pontania*) but a great many more have larvae which feed on herbage. These closely resemble moth caterpillars but have many more legs (or 'prolegs') and have a habit of letting go with

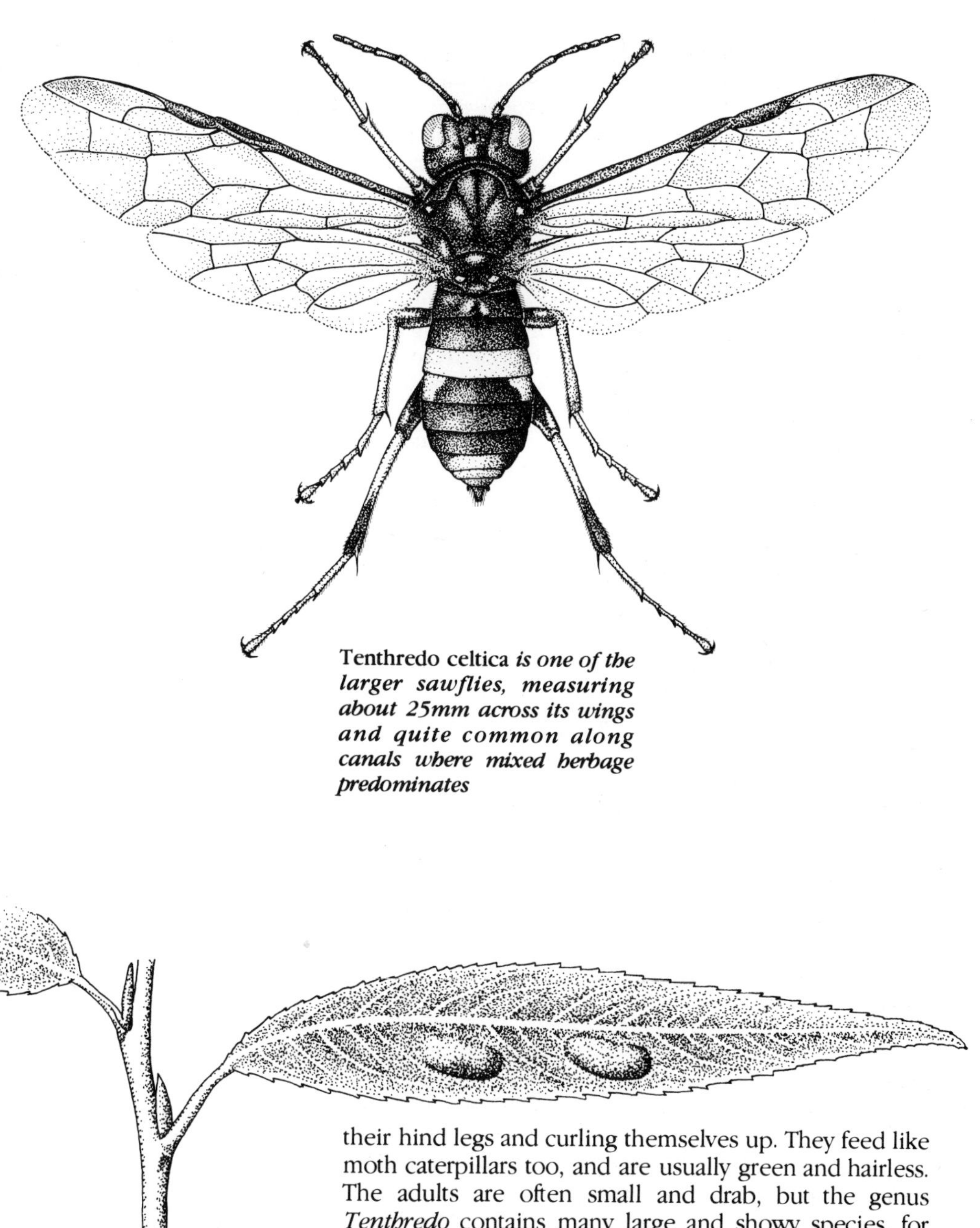

Tenthredo celtica *is one of the larger sawflies, measuring about 25mm across its wings and quite common along canals where mixed herbage predominates*

Bean-galls, made by Pontania *sawflies, are a common sight on willow leaves in the late summer. (10mm)*

their hind legs and curling themselves up. They feed like moth caterpillars too, and are usually green and hairless. The adults are often small and drab, but the genus *Tenthredo* contains many large and showy species, for example *Tenthredo celtica,* a black and yellow insect indigenous to Britain and only recently separated from a continental form. Sawflies fly rather slowly and have opaque wings; they are therefore quite easy to spot but on sunny days are inclined to disappear into herbage to conserve moisture. Flowers are attractive to them, not just for the nectar, but also for the pollen and whatever small insects happen to be around — all are chewed up with equal relish.

Of the more advanced Hymenoptera, only a few of the wasps and bees exhibit a preference for waterside habitats. In the case of bees this is invariably because of a specific association with nectar-rich plants, for example the solitary bee *Macropis europaea* which is very local and is found in certain localities where yellow loosestrife abounds. Wasps are carnivorous and provision their nests with insects, so their distribution is governed by prey species. Thus the wasp *Symmorphus gracilis* is found in association with figwort *Scrophularia* species, because its most regular prey species are the attractive figwort weevils belonging to the genus *Cionus*. Having said this it must be pointed out that figwort is a popular plant with the more usually 'social' wasps too, but in their case they are more interested in extracting nectar for their own energy requirements rather than as food for their young.

As well as the *Cionus* genus of three or four species, figwort provides the home for another interesting beetle, the weevil *Gymnetron beccabungae,* which induces galls in the figwort flowers. It is rather small and does not have the same singular design as the *Cionus* species, and is therefore often overlooked. This also happens to the galls, which are difficult to distinguish from healthy flowers. Of the other riparian beetles, few are dramatic or force themselves to the notice of the general naturalist. The exception in the herb layer is the leaf beetle *Chrysolina menthastri,* which as its name suggests is associated with mint, *Mentha* species, and is a beautiful irridescent green species of medium size which causes the holes in mint leaves but, in its defence, sits in full view like a jewel by the water.

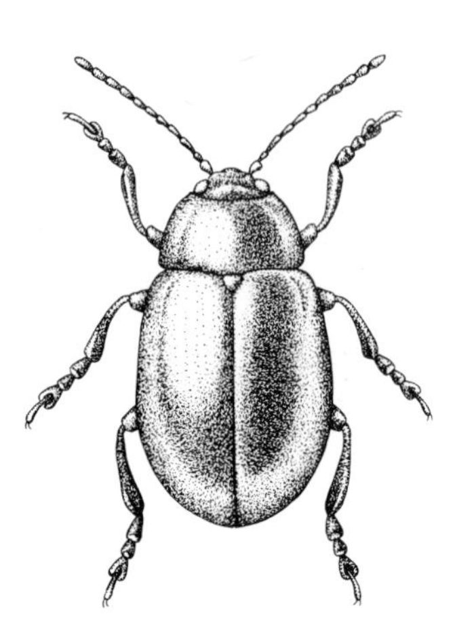

Waterside vegetation is ideal for leaf beetles.* Chrysolina menthastri, *found on water mint, is one of the most beautiful. (8mm)

4 *Shrub Layer*

The different origins of rivers and canals, and thereby the variation in the composition of the tree flora, have a profound influence on invertebrate life. Along rivers, the dominant shrubs like alder *Alnus glutinosa* and crack willow *Salix fragilis* may develop into mature trees and are capable of withstanding severe flooding. They rarely form a continuous barrier, however, and contrast sharply with the planted hawthorn hedges so characteristic of canals. In fact, the best mixture of habitats for invertebrates is that created by disused or rural canals like the Shropshire Union or Kennet and Avon. These have a complete hawthorn hedge infiltrated by other shrub species, with typical 'wetland trees' like sallow wherever a lack of management has allowed them a year or two to establish themselves. In both these canal systems there are continuing schemes to reopen derelict sections but there are still many miles of ideal shrub habitat remaining.

The most abundant of all the riverside trees is the alder, a truly native species and it is difficult to understand why it is

A disused lock on the Kennet and Avon canal in Wiltshire : the grass is flower-rich and the hedgerows are pollution-free

one of the least attractive of all British trees to insects. The only obvious signs of summer damage to its leaves are the 'pouch' galls made by mites of the genus *Eriophyes.* Defoliating caterpillars always leave it alone, and most moths seem to use the trees as an alternative to birch when the latter is not available. This is also true of bugs, beetles and most other insects for which birch is a prime foodplant. Willows, by contrast, are an immensely important family containing such species as common osier, *Salix viminalis,* white willow, *S. alba* and great sallow *S. caprea,* all of which hold rich invertebrate communities. Sallow, in particular, is known for its association with some of our most beautiful and interesting insects. In woodland conditions it is rivalled by the oak, but along waterways it is by far the most important species even though it rarely grows more than 6m (20ft) tall.

Two hawk moths lay their eggs on sallow, the poplar hawk, *Laothoe populi,* and the eyed hawk, *Smerinthus ocellata.* As its name suggests, the former species is also associated with poplar trees, and there seems to be a free interchange between many moths and these two tree groups. Poplars, of course, are closely related to willows and are often found by water.

The caterpillar of the eyed hawk moth, **Smerinthus ocellata** ***is well-camouflaged among sallow leaves. (60mm)***

The eyed hawk caterpillar often gives itself away by its gluttony; it prefers to feed on sallow branches at head height and strips almost every leaf; stalk, stem and all. Its camouflage (an example of 'counter shading') makes it difficult to find, however, since it closely resembles a partly-chewed leaf. Both as a caterpillar and as a moth the eyed hawk is something of an aristocrat, solid yet elegant and never common enough to be taken for granted.

Sallows and poplars are also important foodplants for what are often the favourite moths of moth-hunters, the Notodontids. This family contains the Prominents, an assortment of well-marked and attractive moths with colourful and shapely caterpillars. It also includes the puss-moth *Cerura vinula,* and its acolytes the kittens. The puss has one of the most bizarre caterpillars, apparently designed by a committee to incorporate as many defence mechanisms as possible. If camouflage fails, it has false eye-spots to scare predators, a pair of telescopic whips on its tail to flail at parasites, and a reservoir of formic acid to squirt into the face of unsuspecting humans. A formidable, but irresistably interesting creature. The sallow kitten, *Harpyia furcula,* and the poplar kitten, *Harpyia bifida* are both reasonably common on their respective foodplants, but are not so easy to find. The old cocoon, made up of compacted bark and looking like a nodule low down on the trunk, is often the tell-tale sign. But it is only when the moth has emerged and a hole is visible that it looks anything more than a lump of bark, and fresh cocoons almost always go unnoticed.

Many beetles are to be found on willows. On the leaves, the metallic Chrysomelids are very common, including 'flea' beetles like *Chalcoides aurea;* all are very small but of great beauty. At the other extreme is the musk beetle, *Aromia moschata,* which spends years as a larva burrowing through the wood of old willow trees before emerging as one of the biggest and brightest beetles to be found in this country. The adult is usually bright green and about 3cm long, and gets its name from the sweet smell of musk that it produces. It belongs to the 'longhorn' family, a group characterised by elongated antennae which give its members a primordial quality and must make flight and feeding very complicated.

Willow leaves are attacked not only by moth caterpillars and beetles, but also by sawflies. As has already been explained, sawfly larvae often resemble those of moths; *Cimbex luteus* is one of the largest, being almost the size of a hawkmoth and just as likely to be met with. The most interesting sawflies, however, do not feed on the leaves but cause galls and feed within their protection. The best-known are the 'bean' galls found on willow leaves and caused by *Pontania proxima* and other sawflies of the same genus. The galls are often bright red and appear in the late summer. The sawflies themselves are out in the early spring and can be found on the catkins. Catkins are one of the first sources of nectar and pollen, and consequently a great many other insects visit them too, both by day and night. Butterflies like the small tortoiseshell *Aglais urticae,* moths like the hebrew character *Orthosia gothica,* and a host of solitary bees and flies all rely on sallow blossom to a greater or lesser extent. Of all the waterside trees therefore, the willows are the most essential.

Queens Bower - a beautiful riverside walk through the New Forest in Hampshire

The tradional hawthorn hedges of lowland Britain have always been rich in wildlife. Alongside canals they have become even more prolific, and in total numbers they probably produce even more insects than do willows. The individual species may not always be as outstanding, but the biomass makes up for this and makes the habitat very important for predators. Hawthorn certainly has some very interesting species associated with it, however, including one of the largest British shield-bugs, *Acanthosoma haemorrhoidale,* a green and red creature that feeds on berries in late summer. There are also a great many moths that exhibit a preference for hawthorn; some of the biggest are the Lasiocampidae, a family of fat-bodied, hairy and very excitable animals including the lappet, *Gastropacha quercifolia,* the oak eggar *Lasiocampa quercus,* the pale oak eggar, *Trichiura crataegi* and the small eggar, *Eriogaster lanestris.* Not only are the adults hairy, but the caterpillars are very large with thickly-piled fur that can produce an irritating rash more effectively than a box of itching powder. The lappet caterpillar is second only to the death's head hawk in size, being the same dimensions and colour as a Churchill cigar. The aroma is rather different, but is certainly not unpleasant and has been likened to the smell of rich humus after summer rain. Both the lappet and the oak eggar overwinter as larvae and can be found in the early spring sunning themselves on bare branches before the leaves have opened, but they are well camouflaged then and only half grown.

Young hawthorn leaves are consumed as soon as they open by a whole herd of small caterpillars — especially the winter moth *Operophtera brumata,* the brimstone *Opisthograptis luteolata,* and the attractive (but irritating) yellow-tail *Euproctis similis.* There are, of course, a great many others and it seems incredible that any leaves can

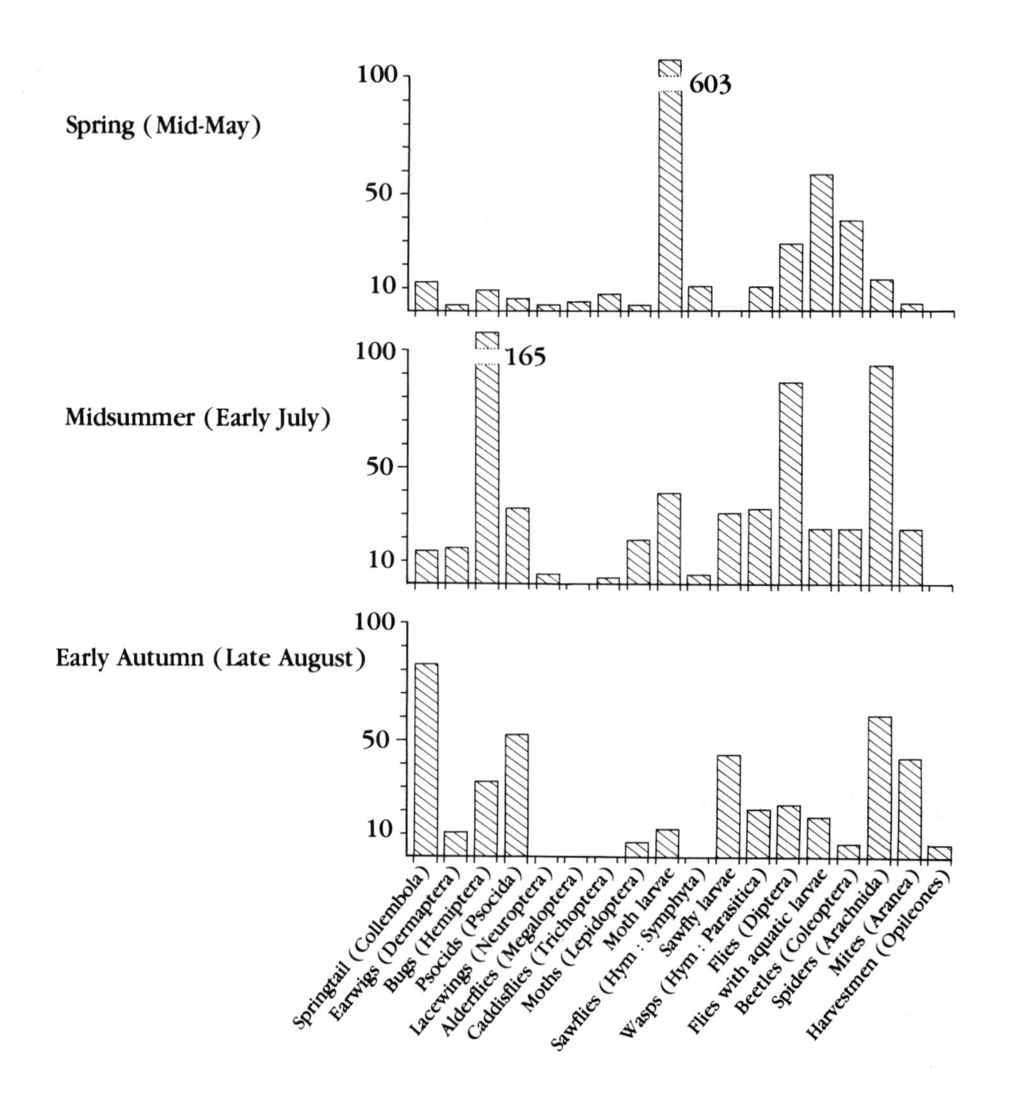

Counts of invertebrate animals beaten from a hawthorn hedge along a 100m section of the Trent-Mersey Canal at Stenson in South Derbyshire. See p158 (The method is simple and provides a useful indication of relative abundance but should not be taken as an accurate measure of total numbers)

survive the onslaught. But the main season of defoliation is short, and the trees are capable of very rapid and continuous recovery.

Hawthorn flowers appear by tradition on the first of May, hence the alternative name of may blossom which has persisted even though a new calendar has made the name obsolete. Their abundance is almost as significant as the earlier flush of sallow catkins, and the swarms of Syrphid hover-flies and *Andrena* bees rely on the supply of nectar and pollen. The blossom is very short-lived, however, and after a few days has completely gone.

The seasonal pattern of hawthorn as an insect resource is much the same as for other trees of course, but along the waterway there is the complication of the emergence and the use of the hedge by semi-aquatic insects like the caddis

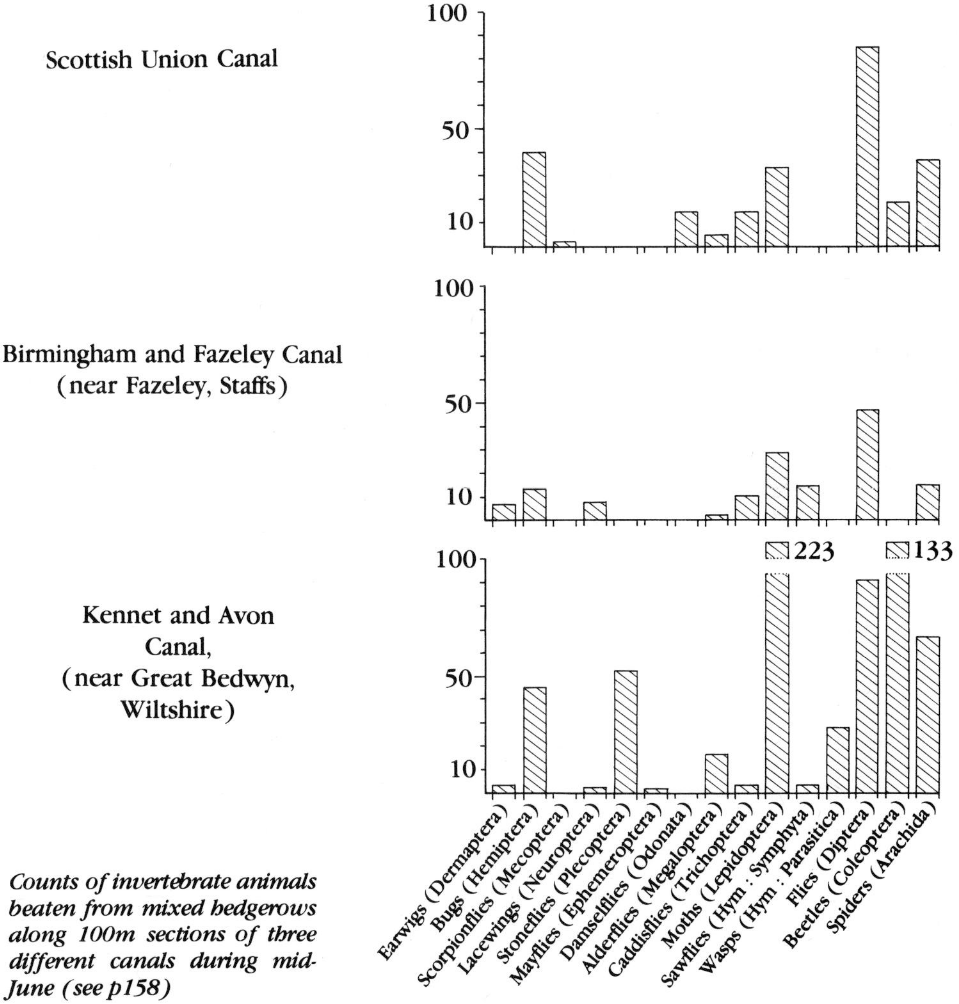

Scottish Union Canal

Birmingham and Fazeley Canal
(near Fazeley, Staffs)

Kennet and Avon Canal,
(near Great Bedwyn, Wiltshire)

100
50
10

223
133

Earwigs (Dermaptera)
Bugs (Hemiptera)
Scorpionflies (Mecoptera)
Lacewings (Neuroptera)
Stoneflies (Plecoptera)
Mayflies (Ephemeroptera)
Damselflies (Odonata)
Alderflies (Megaloptera)
Caddisflies (Trichoptera)
Moths (Lepidoptera)
Sawflies (Hym : Symphyta)
Wasps (Hym : Parasitica)
Flies (Diptera)
Beetles (Coleoptera)
Spiders (Arachida)

Counts of invertebrate animals beaten from mixed hedgerows along 100m sections of three different canals during mid-June *(see p158)*

and alderflies. The richness of canalside hawthorns through the year is demonstrated on page 156. It will be seen from this that most moth larvae make use of the foodplant early in the year and emerge as adults in the mid or late summer, while the opposite is true for sawflies. Also, most other invertebrates reach peak numbers during July or August, but the semi-aquatic species are more closely associated with the spring.

From the entomological viewpoint two factors in particular control the diversity of life on such hedgerows, apart from the purely local considerations of site and aspect. These are: (a) climate, and (b) the quality of the air and water. Page 157 illustrates how significant these can be, using as examples north-facing hawthorn hedges along three contrasting canals. The Kennet and Avon, a clean and underused canal in Wiltshire, is extremely rich, while both the Birmingham and Fazeley (because of pollution) and the Scottish Union (because of its colder climate) are impoverished. Most systems fall somewhere between these extremes and the example provided by the Trent-Mersey on page 156 is probably more typical for the majority of waterways.

In some areas of Britain, the abundance of ash, holly and elm as hedgerow trees at the expense of the hawthorn may actually reduce the range of insects, but naturally-invading trees and shrubs often add to the diversity by providing new foodplants. The addition of an occasional oak to a hawthorn hedge adds greatly to its interest because oak maintains the richest associated community of any British tree. Thus the combination of a hawthorn hedge with shrub oaks and sallows, the proximity of a rich herb layer, and the presence of waterside plants all conspire to make canals one of lowland Britain's few improving resources for invertebrate animals.

11 *Birds and Mammals of the Bank and Hedgerow*

Hedgerows are often thought of as linear woods: the forest edge without the forest, but containing many of the animals that would in nature associate themselves with trees. Riversides often abut woodland or are lined with their own characteristic trees, and canal towpaths have already been shown to contain a vast number of invertebrates on their hawthorn facings. Thus many birds and animals of woods and hedges have adopted the bank systems of canals and rivers without displaying any direct association with water. The habitat provides abundant food: seeds and lush stems for the herbivores and insects

Water meadows and dense riverside herbage are ideal nesting sites for such birds as yellow wagtails and snipe. (River Itchen above Winchester)

The Wye below Hereford : a clean and unspoilt river, rich in birds and waterside mammals

for the insectivores, and contains a variety of breeding or nesting sites. In addition, there are many birds and animals linked with particular vegetation types, and although this association is not as fixed as it is for insects their bias towards waterside habitats makes them easier to see in these situations than in many others.

BIRDS

Finches and Buntings

The scrub zone provides a good crop of seeds, to be eaten during the autumn and winter by finches and buntings that may have bred locally or been attracted over from Scandinavia; the chaffinch and the linnet are the most common finches, but the abundance of welted and spear thistles brings in goldfinches, and alder seeds provide a staple winter food for siskins and redpoll.

The habit of flocking during the winter and autumn is very highly developed in these birds, presumably so that the food supply can be harvested quickly by as large a number as possible without futile searching. Although packs of several thousand (mainly linnets, chaffinches and greenfinches) may feed on adjacent arable land, riverside sites rarely attract more than a hundred at a time. Most search for seeds on the ground, or on the water edge where winter spates may have thrown up a line of flotsam. Siskin and redpoll flocks tend to remain in the tree tops, however, and bullfinches, in small family groups, prefer soft winter buds to hard seeds.

During the spring, finches make extremely neat nests hidden away in bushes and trees, feeding their young on insects and thereby providing enough protein to sustain the necessary rapid growth to fledgling. This habit is shared by the yellowhammer, the commonest of the British buntings, which nests in hedges and is therefore often seen along canal banks. The females can be confused with those of the reed bunting, and the two sometimes feed together during the winter months.

The long-tailed tit, Aegithalos caudatus, *would be one of the smallest British birds if its tail were of normal length. Often overlooked, it is a common woodland species, prefering to stay in small family groups in the treetops*

Tits

The finches' breeding and flocking strategies are shared by the tits, though in a less extreme form. They also feed their young on insects while maintaining a more varied diet themselves, and tend to amalgamate into loosely grouped feeding-flocks in the winter. The mix of species is usually greater, with blue tits and great tits as the staple unit and coal, marsh and willow tits in smaller and less regular numbers. The long-tailed tit, which is one of the most instantly attractive of all British birds, often appears in small family parties with one or two goldcrests in attendance, flitting through the outer branches in search of aphids or other insect eggs. These birds are far more conspicuous when the leaves are off the trees; during the spring and summer they melt into the foliage and are likely to desert the waterside for less disturbed nesting sites.

Thrushes

Hawthorn hedges are favourite nesting sites for blackbirds and song thrushes, which accounts for their abundance along canals. Robins are natural inhabitants of the wood edge too, and are found commonly along canals or rivers where there is sufficient undergrowth to provide a territory. In southern counties, dense cover in damp woodland is the classic habitat for the nightingale, but running water rarely provides the necessary impenetrable tangles.

Warblers

The herb and shrub zones close to the waterside are selected by several species of warblers, often difficult to see and very confusing to identify either by sight or sound. One of the most irritating is the grasshopper warbler which emits a strange reeling song rather like the ratchet of a fishing rod; a continuous and monotonous sound which seems to drift from one place to another even though the bird itself may be stationary. Its habit if skulking in tiny bushes makes it virtually impossible to see unless it is deliberately driven out, and even then it may choose to call the bluff and sit tight.

The marsh warbler, a close relative of the above species but even closer to the reed warbler, is found amongst riverside vegetation in the south-west Midlands. It prefers the herb zone to the emergent vegetation characteristic of the reed warbler and has a very much more attractive song, but remains one of the most difficult birds to identify by sight.

These 'swamp warblers' are joined by four species of true scrub warblers: the whitethroat and the lesser whitethroat, which inhabit tangled briars above the waterline, and the blackcap and the garden warbler, which select bushes or small trees a little farther from the open bank. They are all difficult to see, except for the common whitethroat which usually produces its scratchy song from an exposed perch or even a short song-flight. The blackcap and the garden warbler compensate by their beautiful songs; less powerful than the blackbird and less varied than the nightingale, but very melodic and a great enhancement to what is already a rich habitat.

From the large and confusing group of warblers that flood into Britain for the summer, there remain two very common species to be seen close to rivers and canals. These are the small 'leaf warblers', the chiffchaff and the willow warbler. The chiffchaff is one of the earliest migrants, arriving in March and announcing its presence by its distinctive but limited song. The willow warbler is the most abundant of its family, olive-yellow in colour with an attractive and unassuming song, a descending series of liquid notes, understated and trailing into silence. Both species inhabit woodland but nest near the ground and are often present along waterways where there is birch or tall sallow carr.

Other Insectivores

The yellow wagtail is one of the most beautiful riverine birds feeding and nesting in damp meadowland, but often seen along drainage dykes or river banks. In eastern England, canals figure as an important habitat, but these are usually open-sided rather than the enclosed canals of the Midlands and the south, and must approach the water-meadows that are so much the classic breeding sites. It is a summer visitor, unlike the grey and pied wagtails, and has rather a short tail compared with these species. The male is bright yellow, and the female paler and less eye-catching.

Pipits, closely related to wagtails but resembling miniature thrushes in plumage, are winter visitors to waterways; the meadow pipit is often the commonest of the small birds on muddy riverbanks, having moved from upland moors and rough pasture to make use of the supply of small beetles in flood debris. The water pipit is much rarer and until recently was completely overlooked. It nests on the slopes of European mountains, but appears regularly along English canals and watercress beds; a curious and unexpected contrast and it took bird-watchers many years to catch on to the habit.

Three contrasting species complete the picture of insect-eating birds, the first two are known to everyone and the other hardly at all. The swallow and the wren both feed exclusively on insects, but while the swallow picks up gnats, caddis and stoneflies in open flight above the water, the wren has a systematic strategy based on a minute inspection of every branch along its territory. When all the swallows have left for Africa, the wren remains and usually stays in its regular home throughout the winter. The final species, the dunnock, shares this characteristic but switches to eating seeds or berries, employing much the same technique as the robin. During very severe weather, when canals and even rivers may be frozen, large numbers of wrens may die but the dunnock usually manages to survive.

Crows

The magpie and the carrion crow are the scavengers of riverbanks, picking over the leftovers when carnivores like foxes or mink have finished, and tackling small prey themselves when there is nothing else available. Where waterways cut through woodland, the jay joins them but is much more retiring and is usually heard rather than seen as it disappears among the trees. Its obsession with acorns in the autumn often draws it out into the open; its colours make it one of the most gaudy birds, yet at a distance it can merge into the background surprisingly well.

Birds of Prey

Most birds of prey are opportunists, and will take whatever food presents itself. There are records of both tawny owls and barn owls snatching fish from water, but in general anything beneath the surface is out of bounds. The exception of course is provided by the osprey, and although this is still an uncommon bird, nesting on Scottish lochs rather than rivers, it migrates south each autumn and is seen in many unexpected places taking fish from canals or village ponds as well as open rivers and lakes.

Hedgerows are a traditional hunting site for both the kestrel and the sparrow hawk, the former mainly in search of voles and the latter of birds roosting in the branches. Since they are the commonest diurnal birds of prey they are most often seen along waterways, making use of the associated habitats.

MAMMALS

Insectivores

Only one of the insect-eating mammals hibernates, the rest find enough food in leaf-litter or the soil to keep themselves going throughout even the coldest weather.

This group includes three shrews, of which the water shrew has already been considered, together with the mole and the hedgehog. The hedgehog, of course, is the species to hibernate and it does this between October and April — though some young animals may be about until December. During the summer the hedgehog is a common visitor to the waterside at night, the delicate footprints appearing in meandering lines over long sections of river mud, suggesting that beetles and grubs are readily available. The eggs of ground-nesting birds are a particular delicacy, and moorhens suffer particularly from the predation by hedgehogs if the nests are at all accessible.

The mole is a much more specialised animal, feeding almost entirely on worms, but occurring wherever there are a few centimetres of soil. In the rich, deep alluvium alongside rivers it is an abundant animal where there is little risk of flooding, but it is also found along canal banks where tell-tale hills may be hidden by the leaf-litter. The tunnel system is complex and built on different levels, each series containing only a single mole. Once excavation is complete there is no need for more hills, but during very cold weather the upper surface of the ground may freeze, forcing the mole to follow the worms lower down. Flooding is a serious problem, so the inhabitants of river plains are particularly at risk. If there is little warning of a coming deluge moles may easily drown in their burrows, but since they are accomplished swimmers they are more likely to be able to swim to safety. They will, of course, be faced then with the problem of trying to stake out a territory in an area already occupied, for moles are extremely aggressive to each other and confrontations are likely to be explosive.

As well as their normal hills, moles sometimes produce 'fortresses'; structures of a metre high which often contain nests and are far more elaborate than the normal mound.

These may be built to provide refuge above flood level and are quoted as a sign of bad weather to come, but they are also found on downland and so this cannot be the only explanation. It seems likely that fortresses are built for a variety of reasons, dictated by circumstances and the inclinations of the mole concerned.

The common and the pigmy shrew both eat worms too, but they will also tackle a variety of invertebrates, accounting for just less than their own body weight in food every day. They are tiny all-action animals constantly on the move, and are most common among grassland or hedgerows. For this reason, and because they are active during the day as well as at night, they are regularly seen along canal towpaths. Their dead bodies are often seen too; the long pointed nose, tiny eyes and grey fur are sufficient to distinguish them from rodents, and the red-tipped teeth do not have the broad yellow incisors of mice and voles.

Many bats hunt over water because of the abundant supply of insects. Daubenton's bat,* Myotis daubentoni *is the only species particularly associated with the habitat

Bats

Writing in 1767, Gilbert White noted the abundance of bats between Richmond and Sunbury, so much that 'the air swarmed with them all along the Thames'. Although several species of bat hunt insects over the water, it is the Daubenton's that has gained the reputation as a specialist to the extent that it is sometimes known as the water bat. It is not, however, confined to this habitat and will hunt woodland just as freely.

Identifying bats in flight is virtually impossible unless a bat detector (recording the different ultrasonic calls) is used. The pipistrelle, which is another regular waterside feeder, measures only 20cm (8in) across its extended wings, while Daubenton's is about 25cm (10in). Apart from this, Daubenton's appears rather heavier with broader wings, but on a dark night it is unlikely that such subtleties will be very much use. Both species roost communally in hollow trees, but the Daubenton's moves off in September to hibernate in caves or in old buildings.

Of the other British species, the most likely to be seen are the long-eared and the noctule — the latter being the biggest with a wingspan of up to 40cm (20in). Both are strongly associated with trees and are common along wooded valleys.

Day-flying moths are often mistaken for butterflies, especially if they are colourful. The scarlet tiger, Callimorpha dominula, *is one of the most attractive* ▷

Rodents

The bank vole is the most likely small mammal to be seen on riverside footpaths; it is abundant among tall herbage and hedgerows and is partly diurnal. It is very much smaller than the water vole and the coat is a rich red-brown, which makes it a distinctive species and easy to identify. The wood mouse is equally common, but it is a much more agile creature and is unlikely to be seen in open daylight. The most obvious difference between mice and voles is the length of the ears and tail; mice have large ears and very long tails, while voles do not.

Of the larger rodents, the rats are by far the most unsavoury animals of the waterway. There are two British species both introduced accidentally from shipping and both from Asia. The first to arrive here was the black rat which was responsible for the spread of the Great Plague, but it is now confined to sea ports where it is an urban specialist. The brown rat is far more successful and has developed a liking for sewers and derelict urban canals. In the process it has become confused with the water vole but it is readily identified by its pointed nose and long tail. It can swim almost as well as the water vole and does so without hesitation.

Both of the British squirrels, the red and the grey, visit river banks and their footprints can often be seen some distance out from the bank when rivers are frozen and food may be available on the ice. The grey squirrel is the only species encountered in most of lowland England; the red has retreated to the coniferous woodland of the north and west.

The mottled green underside of the orange tip butterfly, Anthocharis cardamines, *serves as camouflage among low herbage* ▷

Carnivores

The fox, the badger, stoat and weasel are all important predators of riverside wildlife. The nests of ground or shrub-nesting birds are heavily predated by stoats and adult ducks form a regular component of the fox's diet. All these hunters use river and canal trackways and their footprints and droppings can be found as regularly as those of the true river specialists. The polecat, which is spreading from its stronghold in Wales and the Marches, also displays a wide tolerance of habitat-types, and is apparently quite common along the upper reaches of the Severn and along the Shropshire Union Canal. It is intermediate in size between the stoat and the mink and, together with its domestic form the ferret, may be introduced either deliberately or accidentally into other parts of Britain.

The Future

12 *Conservation Problems*

The main threat to our wetlands is drainage, the aim being to improve agricultural land, so that arable crops can be grown. The lower agricultural value of wetlands also makes them prime candidates for tipping and industrial and housing development. However, a more insidious cause of loss of our wetlands is the lowering of the water table by water abstraction, which is especially significant in lowland areas adjacent to centres of population, where the rainfall is low. Changes in plants in chalk streams have already been noted, with a loss of the submerged plants and the habitat they provide. The use of low-lying marshes and flood plain areas for building increases the risk of flooding; this leads to flood prevention works which alter the character of the river and drain further wetlands.

In 1977 a *Red Data Book of British Flowering Plants and Ferns* was produced, listing all threatened species. By 1981 63 per cent of all the wetland species in the book were either extinct or under severe threat of extinction. The loss of ponds is high; in Nottinghamshire as many as 90 per cent have gone since 1945; marshes are less easy to trace on maps but they too have disappeared at an alarming rate.

Canals, rivers and their environs, while not providing all variations on the wetland theme are important for many pond and marsh species. Even so, there are threats to waterways especially as a result of river management, pollution and recreation.

River Management

There are two types of river management, large scale alterations, usually called improvements, which are designed to alleviate flooding or improve drainage, and secondly, the maintenance works. Improvements basically involve the removal of variation by straightening and deepening the river, smoothing the bed and often creating steep embankments to contain floodwater; the aim is to remove water quickly from areas where it causes problems. The pools, meanders and shallows that are important for some fish are replaced by a graded bed

◁ ***A fully-grown caterpillar of the elephant hawk moth,*** **Deilephila elpenor**, ***feeding on orange balsam***

which is unsuitable for them. Weirs without fish ladders may prevent migratory fish from reaching spawning grounds. The absence of a shallow shelf and marginal vegetation discourages herons, grey wagtails, green sandpipers and similar birds.

The straightening evens out the flow, and plants find it difficult to recolonise and therefore the animals they shelter are lost too; the graded bed is often unstable and this also hinders the return of the wildlife. The new sloping embankments rarely replace the sand martin breeding banks, particularly as they are usually sown with an agricultural seed mix and mown regularly.

It must be admitted that some trees become unstable but often they are removed for convenience to allow easy access for machinery, yet the fibrous roots of willows and alder help to hold a bank together. The removal of trees and shrubs eliminates food and shelter for many birds and invertebrates and also small but significant things like kingfisher perches. Once light has been let in the plant communities change, the growth on the river bed and banks increases and reed sweet-grass, *Glyceria maxima,* bur-reed, *Sparganium emersum* and fool's watercress, *Apium nodiflorum,* come in. Brambles, *Rubus* species and perennial grasses dominate the banks, and the variety of plants and animals diminishes.

Improvements to river banks often remove trees and shrubs which are vital to many foraging insects. The solitary wasp* Ectemnius lapidarius *nests in old beetle holes in rotten wood, but such sites are often cleared

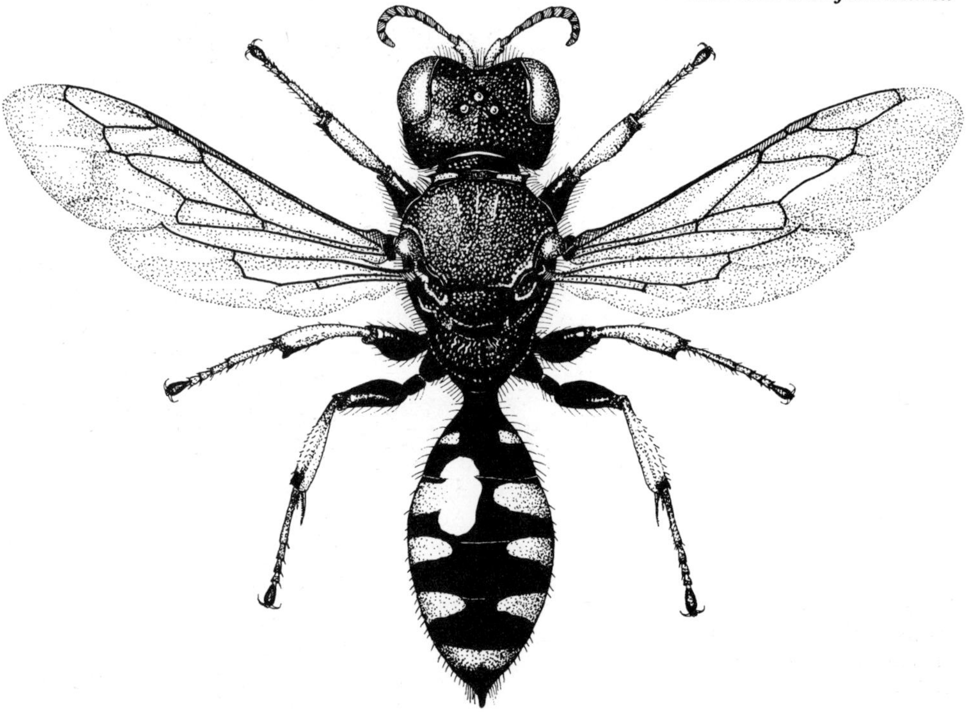

Conservation Guidelines
However, most Water Authorities, who have the responsibility for much of the work are altering their methods of working after pressure from various conservation bodies, particularly the Royal Society for Nature Conservation and the Royal Society for the Protection of Birds. In 1980, the Water Space Amenity Commission published *Conservation and Land Drainage Guidelines* which indicate ways of reducing damage to wildlife communities. Some Water Authorities ask the County Conservation Trusts to do surveys of rivers before detailed plans are made, identifying important stretches and features. While it is not possible to retain all of them, the situation is improving as understanding on both sides grows. The machines now work from one side only, perhaps alternating along a stretch, which allows trees to be left on the other side; some are pollarded during the work to enable machinery to work but they will regrow afterwards. Landowners may be approached about replacement tree planting. Fishing interests are often allies because they want the maintenance or reinstatement of the diversity necessary for the feeding and breeding sites and

A pollarded willow : a traditional method of wood production but the tree is deformed in the process

Sluice gates on the Great Ouse at Denver in Norfolk. The artificial control of water for land drainage, water supply and amenity all affect the natural flow of rivers

shelter which the fish require. However, any work disrupts the communities and recolonisation still depends on the seeds, plant fragments and invertebrates arriving from untouched tributaries upstream, although the right conditions may take time to reappear. The other problem is that once a natural river has been interfered with, maintenance, dredging and weed control, often becomes necessary.

Maintenance Work

Canals and dykes are artificial and maintenance, such as dredging to remove accumulated silt to let boats or water pass also maintains the habitat diversity. Although the early stages of disuse are rich in wildlife the small gradient encourages siltation which results eventually in a habitat poor in species often dominated by reed sweet-grass, *Glyceria maxima.* If dykes are cleaned out in stretches on a rotation, to allow recolonisation, and only the canal channel is dredged, leaving marginal vegetation, recovery is usually good.

In some southern rivers (for example the Itchen) vegetation is cut using hand tools or weedcutting boats, to improve the habitat for game fish and to remove weed liked by the 'undesirable' coarse fish. In canalised rivers and dykes vegetation may impede the flow; in canals, since propeller driven boats were introduced, cutting is no longer necessary to clear the channel. The breakdown of the plant material resulting from cutting or the use of herbicides deoxygenates the water at a time (summer) when temperature and other factors are also reducing the supply. The removal of one group of plants, for example emergents, results in the dominance of another, which may be worse. Algal blooms develop in unstable systems and add to the oxygen shortage.

Indirect effects on aquatic communities are caused by water abstraction and the building of reservoirs both of which even out the flow. Abstraction may also reduce the flow to a point where salmon and trout cannot swim and exaggeration of pollution and temperature effects occurs. Another newer factor is water transfer across watersheds which may spread species like the canals did; it also results in interference with natural flow patterns which are so important in the ecology of waterway species.

Pollution

Waterways have long been used as sewers and dustbins for the toxic products of industrial processes. In 1972 2,000km (1,240 miles) out of 32,000km (19,900 miles) of waterways were still 'grossly polluted', but it is true that most are cleaner than they have been for a long time. The Rivers Tame and Thames are the most quoted examples; the Thames was without fish as recently as the late 1950s, but now over forty species have been recorded. Canals and lowland rivers suffer most from pollution because they pass through more cities and towns, the flow is slower and smaller and consequently dilution is less.

The industrial and domestic wastes that causes pollution have been largely controlled but agriculture poses new threats, from fertiliser run-off, pesticides and the slurry from large scale animal units. The main effects of pollution are the sediment, acute toxicity and, indirectly, deoxygenation.

Poorly treated sewage (from overloaded sewage works) may result in fine particles in the effluent, but so too do sand and gravel washings, soot, coal dust, ochre (from mine washings) and china and ball clay, if not adequately filtered or settled. The particles cloud the water and reduce light, they clothe plants and make it impossible for algae and invertebrates to attach themselves and they clog animals' gills and plant pores.

Toxic chemicals include pesticides as well as ammonia, copper (which kills algae), chromium and cyanide from paper making, metal working and chemical processes. Fish may die in large numbers and even if they are restocked the rest of the community will take longer to return and the substance may stay in the sediments and be disturbed years later by dredging. Road accidents are an important cause of pollution, while many factories still do not have adequate treatment plants. Guidelines for the use of pesticides can be ignored causing spray drift in high winds, the recommended dose can be exceeeded and spraying in the rain helps no-one! Fish concentrate organochlorine pesticides such as DDT and dieldrin to lethal levels and may be eaten by herons which then suffer sub-lethal effects for example poor breeding success. (See also the section on the otter in chapter 8).

In the third chapter, the importance of oxygen was stressed; unlike air, water can lose all its oxygen. Organic matter, such as partly treated sewage, slurry from farms (silage and manure) and effluent from breweries and milk processing factories, is attacked by decomposers, bacteria and fungi, which use oxygen. The amount of oxygen absorbed by a given amount of polluted water is known as the Biological Oxygen Demand (BOD); the greater the BOD the greater the amount of pollution. Aeration from locks, weirs and cooling towers is very important particularly since the sewage from one person will require the oxygen from more than ten thousand litres (2,000gal) of water each day. Progress has been made towards secondary treatment, reducing the BOD of effluents.

Hard detergent 'swans' (fluffy masses of foam) which stopped oxygen and light entering the water are a thing of the past, but phosphates from modern detergents do pass through sewage works and with nitrates from fertilisers and other nutrients from farm wastes produce artificial eutrophication. Fertiliser usage has increased nearly five-fold since 1945 and it is estimated that about half is wasted, some of which leaches into our waterways. The problem is made worse by the increasing use of liquid ammonia instead of insoluble nitrates. Aquatic and emergent vegetation often grows well after rain in May, just after the spring fertiliser application. Later on algal blooms develop, blotting out light, tying up nutrients, using oxygen during

The Warwick and Napton Canal (now part of the Grand Union). A large patch of arrowhead in the centre of the picture has escaped the effects of leisure craft

life and in decay and often preventing oxygenation of the water causing 'fish kills'. Thermal pollution and its effects have been dealt with on page 37. .

Although rubbish may be unsightly, inert items like wooden pallets and polystyrene blocks may provide new habitats. In other cases it is a source of pollution when cans and drums containing chemicals or oil are thrown in. Debris on the bottom such as bicycles, prams and supermarket trollies may seem innocuous enough, creating slight stagnation and an unsightly nuisance, but mud and vegetation are stirred up as boats try to get round the obstruction and the effect on the habitat may soon become permanent.

Recreation

Water holds a great attraction for many, for an afternoon walk, a boating holiday or fishing. Even canal restoration is a popular recreation. Each has an effect on the wildlife, either temporary or permanent, whether it involves trampling the banks or disturbing birds; and research has already been carried out on many of the problems. Plants and animals may be physically damaged or affected by environmental changes and animals also react to the presence of people, the noise they create and alterations in the plant communities.

Boating

Boating for pleasure began late in the last century and *Salter's Guide to the Thames,* published in 1894, included descriptions of southern and Midland canals. The main craft then were the canoes, punts and skiffs of Jerome K. Jerome's *Three Men in a Boat.* Steam power was developed in 1793, but it was not widely used, mainly because of the effects of the wash; the real changes only started in the 1930s when oil engines superceded horses on most commercial craft. Although the first cruiser hire firm was established (near Chester) in the mid-1930s, there was little expansion until thirty years later. In 1968 there were approximately 8,000 privately owned powered boats and 1,000 available for hire on the rivers and canals of Britain. Ten years later there were over 20,000, including 2,500 hire boats. In 1977 British Waterways Board announced that recreational boating had exceeeded the use made of canals when they were solely commercial. This increase in use, imposed on a system which had not been in full use for sometime, has created problems of maintenance as well as for wildlife. Recreational boating reaches a peak in summer, which puts a strain on a system designed to cope with regular, year round traffic.

The main effects of boating are the physical damage caused by propellors and by the boats themselves, the wash and turbidity. All boating activities are potentially

The River Thames at Wallingford in Oxfordshire, where amenity pressures are particularly high

disturbing to some groups of animals, but research so far has concentrated on wildfowl and fish.

The cutting action of propellors kills some plants, although others can regrow from the fragments. The draught of most pleasure boats is considerably less than that of working narrowboats and barges and therefore they can get in amongst the marginal habitats in 0.3-0.6m (1-2ft) of water where submerged, floating and emergent plants grow. Waterlilies with their long leaf-stems suffer greatly and often only the cabbage-like submerged leaves are left in busy canals by mid-summer. Boats crush plants in locks and when they are mooring; once tied up the movement of the boat as others pass enlarges the gap. Working boats tended to moor, except in emergencies, in regular mooring places whereas holidaymakers affect a greater length of banks by their more haphazard choice of stopping places.

The effect of wash on banks: a bow wave which erodes the banks and severely disturbs emergent and aquatic plants

There are speed limits on canals and most rivers, but even if they are obeyed wash creates movement at the edges and on the bed, unless the waterway is wide or deep. As the boat moves the water level at the edges ahead of it rises, then drops quickly as it passes, leaving a series of waves, exerting a considerable strain on plants from all sides. Wash may overturn floating leaves, submerging the pores, or it can cause a loss of marginal vegetation, which is pulled out or washed away on busy stretches, leading to bank erosion; the remedy for this, piling, causes further disruption to wildlife. Submerged, floating and emergent plants are all affected, although those without dense roots are more susceptible. The effects of wash on animals are largely unknown but birds' nests are swamped, emerging dragonfly nymphs are washed away and other aquatic animals are stranded.

Busy waterways, such as the Grand Union Canal rarely have clear water, even in winter when traffic is reduced, which restricts both plants and animals. Powered boats are also a potential pollution hazard. Bilge water from inboard engines can have a lot of oil in it and two-stroke outboard engines may lose 10-20 per cent of the fuel used into the water. Unpowered craft, particularly canoes, are increasingly used on waterways. Rivers offer more excitement for canoeists, but canals provide a good training ground for young people. Sailing and rowing are restricted to reservoirs, for instance the Welsh Harp in Brent and wider rivers such as the Trent and Thames.

Regular launching areas may be trampled or concreted and canoes, sailing and rowing boats are shallow enough to reach parts which are normally avoided by cruisers and narrowboats. While the wash they create is not very strong, it can remove vegetation like shoreweed, *Littorella uniflora,* from shallow margins and oars and paddles cause damage too. Studies have shown that bottom invertebrates have disappeared from shallow rivers used for canoeing and fish breeding also appears to be affected.

Angling

Angling is found on every type of water, non-navigable and navigable rivers and canals, canal and water supply reservoirs and on the coast. Nearly three million people participate in the sport, with coarse fishing more popular than sea-angling, and game fishing, the most expensive, coming a poor third. Many anglers, based in the Midlands, fish locally for coarse fish and make the occasional trip to the coast. About 500 miles of waterways are not fished because they are too polluted, but the rest is fished by individual riparian owners, by club members who have leased banks or feeder reservoirs or by non-members with day-permits for certain sections. Clearly some parts of the system, near to cities and towns are heavily fished and the remote sections receive less pressure, but nowadays anglers are able and willing to travel large distances for their sport.

The chalk-rivers of southern England are managed for fly-fishing but are extremely rich in wildlife. A side-stream of the Test at Leckford in Hampshire

One of the most obvious effects of fishing is the clearance of bank and marginal vegetation for a 'peg' to gain access to the water. The vegetation is often cut early in the season and with subsequent trampling perennial rye grass, *Lolium perenne,* meadowgrass, *Poa pratensis,* rat's tail plantain, *Plantago major,* and other low-growing plants replace the tall bank community. On the River Ouse near Huntingdon, 30 per cent of the bank vegetation has been altered in this way near access points. Marginal plants are cleared for keep nets and casting and in weedy areas aquatic vegetation is also removed, all fragmenting the habitats. Reedbeds are vulnerable to trampling and may be destroyed, yet fringing reed margins are desirable for fisheries, sheltering brown trout and pike.

Hook, Line and Sinker

Discarded hooks and line kill or injure thousands of birds every year, ranging from coots and mallards to sparrows caught up in hedges nearby. There is a great deal of evidence that the split lead shot used to weight lines in coarse fishing is the cause of lead poisoning in waterfowl. A report published by the Nature Conservancy Council in 1981 *Lead Poisoning in Swans,* estimates that 250 tons of lead in the form of shot is introduced into the environment each year. This is equivalent to two pieces of shot per foot per year along all lowland rivers and canals. Mute swans (and some other waterfowl)feed mainly on submerged aquatic vegetation and need grit to aid digestion. Discarded split lead shot on banks or in the water may be mistaken for grit and once swallowed affects the nervous system, preventing the normal working of the gizzard, and eventually causing death by starvation as well as by poisoning. The problem is particularly acute on certain

Hook, line and possibly sinker! . . . A mallard saved from an unpleasant death

rivers, notably the Trent, Avon and Thames, where there has been a marked decline in the number of swans over the past ten to fifteen years. On these rivers between 1973 and 1981, 75-90 per cent of swan deaths were due to lead poisoning. By contrast very few cases of lead poisoning are recorded in Scotland where game fishing is dominant. The report estimates that in England alone between 2,700 and 3,500 swans may be dying each year from lead poisoning. Swans are distinctive, large birds and it is possible that the deaths of other waterfowl, which feed in the same way, go unnoticed.

Another impact of fishing is on the fish themselves. Fish are hooked, handled and put in keep-nets causing loss of scales and other damage which reduces their chances of survival. The transfer of native fish and the introduction of alien species by angling clubs can upset an existing ecological balance and can also introduce disease.

Walking

The legal position of walkers is complicated because few canals and fewer rivers have a continuous public right of way. Canal tow paths were made under the enabling Act of Parliament for the use of the animals and men who towed the boats. Until 1978 the British Waterways Board issued a special permit for walkers using tow paths; since then the public have permission but not a right to use the paths and they have to take them as they find them because the Board is not responsible for tow paths. Walkers, including naturalists, trample and disturb wildlife, just like other users and the increase in accessibility to rivers and canals is thought to be an important factor in the decline of the otter (see pages 121-2).

Pollution, management and canal restoration affect large areas of habitat whereas recreation tends to affect individual animals or plants and only where pressure is continuous or widespread are species likely to be threatened. However, with the loss of wetland habitats elsewhere, and the likely increase in recreation on or by the water the problems of conserving the variety of plant and animal life will grow; moreover the gaps in our knowledge of the existing situation and of the long term effects of recreation leave us ill-prepared to identify any changes. It is to be hoped that those who sail, cruise, fish or walk along our waterways will take notice of the wildlife that forms the backdrop of their activities and not just mourn its passing when it is too late to conserve it.

13 *Things to Do/ Places to Go*

The advantage of studying river or canal habitats is that they are varied and accessible, and include a body of water rich in unusual life-forms. Children always enjoy pond-dipping but it should be remembered that canals and lowland rivers are rarely safe. Studies of canals are hampered by the confusing variety of invertebrates, but there are text-books available and the equipment necessary is not expensive; a strong net with a fine (1mm) mesh for free-swimming creatures and a kitchen sieve for the muddy substrate. An enamel or plastic dish should be used to empty the catch into, and specimens should be identified on the spot and released wherever possible.

Children always enjoy pond-dipping : the equipment is comparatively simple and the results are nearly always dramatic

Sampling insect life on riverside vegetation, using a sweep net. This is a convenient and productive way of finding the less obvious insect groups such as caddis and stoneflies

There is nothing wrong with taking some creatures home for study in an aquarium, but less common creatures should be returned to the water immediately.

The invertebrates of emergent and waterside vegetation are caught using either a black cotton butterfly net of fine mesh (for flying insects) or a sweep net of strong white calico. The latter is drawn through the herbage and is very efficient at obtaining large samples of insects and spiders, but it can damage plant life and should be used sparingly. For trees and high herbage, a beating tray is a useful item — a large white tray onto which invertebrates fall after having been dislodged from the branches. None of the above equipment is essential, however, and a good pair of eyes is the most valuable aid in field studies work.

For identifying birds a pair of binoculars is very nearly indispensible. 7x50 is the ideal specification for riverside work because the magnification is not too great; small warblers can be lost among the reeds if the field of vision is too restricted. Mammals are a different proposition, and it is unlikely that they will be seen at all unless a particular effort is made at dawn or dusk. The study of tracks and signs makes up for this to some extent, particularly as a collection of plaster casts can be made during the winter months when other wildlife activity is at a low ebb.

The identification and interpretation of tracks along a river-bank can be both challenging and interesting. Here, the distinctive footprints of a water vole are seen crossing an area of cracking mud. Before this happened, and before the mud was uncovered, a heron had been walking in the shallows (footprints mid left of photograph). The heavy boot-print of an angler, made in the early morning after the vole had passed, can be seen to the mid right.

Publications on river and canal habitats are very limited, but identification guides are only a problem for the more obscure invertebrate groups. The following list provides a cross-section of worthwhile books still in print.

Pond Life (Young Specialist Series), W. Engelhardt, published by Burke — simple but well illustrated.

Freshwater Life, John Clegg, published by Warne — still the standard text book, authoritative but not well illustrated.

Rivers, Lakes and Marshes, Brian Whitton, published by Hodder and Stoughton — an attractive book with excellent introductory essays, but of limited value as an identification guide.

Field Guide to the Insects of Britain and Northern Europe, Michael Chinery, published by Collins — the only attempt to survey all the insect families, with identification down to most genera. Very well illustrated and an essential first guide.

The Dragonflies of Great Britain and Ireland, Cyril Hammond, published by Curwen — an example of what should be done for all insect groups. A classic book, and easy to use.

Key to the Fishes of Northern Europe, Alwyne Wheeler, published by Warne — an excellent text, illustrated by line drawings.

Guide to the Freshwater Fishes of Britain and Europe, B.J. Muus and P. Dahlstrøm, published by Collins — illustrated in colour, not quite such a good text.

The British Amphibians and Reptiles, Malcolm Smith, published by Collins — the best book on the subject ever written.

Field Guide to the Birds of Britain and Europe, R.T. Peterson *et al,* published by Collins — the first of a whole galaxy of bird identification books, and still considered the best single book, though the illustrations have been superceded.

Mammals of Britain; Their Tracks, Trails and Signs, M.J. Lawrence and R.W. Brown, published by Blandford — good for details of tracks with many drawings.

As with animals, there is a need to consider the conservation of plants and it should be remembered that it is illegal to uproot any wild plant, without the permission of the landowner. It is also undesirable to pick wildflowers, because this prevents seed production and therefore the book should be taken to the plant. In the case of water plants it may be necessary to take a fragment to study it, but always return it to the water, where it may regrow.

The identification of flowering plants presents fewer practical problems than that of animals, but the variety of identification books can be confusing. Every amateur botanist has their favourite identification book but for the beginner, the *Collins Pocket Guide to Wild Flowers* by D. McClintock and R.S.R. Fitter is probably the best, because the illustrations are arranged according to the colour of the flower and the descriptions are good. The disadvantage of this book, that flowers are only present for part of the year, is avoided by *The Wild Flower Key* by Francis Rose, published by Warne, which has a key based on features other than flowers. It also introduces the idea of the family groupings of plants, which is an important step and one that is necessary when using *The Concise British Flora in Colour* by W. Keble Martin, published by Michael Joseph, which has the best illustrations. *British Water Plants* by S. Haslam, C. Sinker and P. Wolseley, a reprint from the *Journal of the Field Studies Council*, uses vegetative (non-flowering) characteristics and has clear illustrations.

Non-flowering plants such as mosses, liverworts, lichens, horsetails, ferns and fungi are dealt with, grouped in habitats, in an excellent book, *The Oxford Book of Flowerless Plants* by F.H. Brightman and B.E. Nicholson. *A Beginners Guide to Freshwater Algae*, published by HMSO is very clear and certainly whets the appetite for a group that is overlooked.

Many specialist books exist for most of the interesting animal and plant groups and the national societies such as the Freshwater Biological Association (The Ferry House, Ambleside, Cumbria)and the Botanical Society of the British Isles (Botany Department, British Museum (Natural History) Cromwell Road, London SW7 5BD) publish scientific keys and offer help to the more committed naturalists. For the not-so-committed, membership of a local county naturalist or conservation trust (addresses from the Royal Society for Nature Conservation, The Green, Nettleham, Lincoln LN2 2NR) will provide not only contacts with other interested people, both beginners and experts, and opportunities for field trips, talks etc, but a chance to help conserve the habitats as well.

The Thames at Wallingford, Oxon

NATURE TRAIL GUIDES

The following is a list of trail guides for various canals and rivers. Although many waterways societies, all the County Conservation Trusts, the Royal Society for the Protection of Birds and many other organisations have been contacted, it is unlikely to be a full list, but it should provide a starting point. Some of the publications only mention wildlife in passing, others are full nature trails.

River Almond, Midlothian
Almond Valley Nature Trail, from the Information Office, Midlothian County Council, George IV Bridge, Edinburgh 1.

Ashton Canal
Brabyns Park Interest Trail, from East Area Office, Administrative Division, Metropolitan Borough of Stockport, Memorial Park, Marple, Stockport, Cheshire SK6 6BB.

River Avon
The Wildlife of the Higher Avon and it Future – Interim Report, from Warwickshire Nature Conservation Trust Limited, 1 Northgate Street, Warwick CV34 4SP.

Basingstoke Canal
Natural History of the Basingstoke Canal, from Surrey and Hampshire Canal Society, 51 Elmsleigh Road, Farnborough, Hants.

Old and New Bedford Rivers - Ouse Washes
Ouse Washes, from the Cambridgeshire and Isle of Ely Naturalists' Trust, 1 Brookside, Cambridge CB2 1JF.

Birmingham Canal Navigations
BCN Main Line; Smethwick Town Trail, from local libraries.
Wyrley and Essington Canal; Rough Wood Nature Trails; Daw End Branch Canal; Hay Head Nature Trail, from Walsall Borough Recreation and Amenities Department, Gorway Buildings, Gorway Road, Walsall, West Midlands.

Bollin River
Bollin Valley Teachers' Pack, from Cheshire County Council, County Hall, Chester CH1 1SF.

Bridgewater Canal
Bollin Valley Teachers' Pack, from Cheshire County Council, County Hall, Chester CH1 1SF.

Bridgwater and Taunton Canal
Severnside: A Guide to Family Walks, includes a walk along the canal. Edited by Stephen Taylor; published by Croom Helm.

The Broads

What to do on the Norfolk Broads, published by Jarrolds; *The Broads Book* published by Link House. Both of these books are produced each year and contain information on nature reserves and trails.

Hoveton Great Broad Nature Trail, from Nature Conservancy Council, 60 Bracondale, Norwich.

Ranworth Marshes Nature Trail, This leads to the Norfolk Naturalists Trust Broadland Conservation Centre. Details from the Norfolk Naturalists Trust, 72 Cathedral Close, Norwich, Norfolk. The Trust also operates various boat tours and trails at Hickling Broad National Nature Reserve.

Other information is available from the Broads Authority, Thorpe Lodge, Yarmouth Road, Norwich NR7 0DU.

Bude Canal

Tamar Lake (a supply reservoir for the canal) in *Devon Wetlands,* published by Devon County Council, Topsham Road, Exeter.

Calder and Hebble Navigation

The Calderdale Way, information from South Pennines Information Service, 1 Bridge Gate, Hebden Bridge HX7 1JP.

Caldon Canal

The Caldon Canal, from the Caldon Canal Society, Canal Basin, Foxt Road, Froghall, Stoke-on-Trent.

Staffordshire Way – Part 1, from Chief Planning Officer, Staffordshire Moorlands District Council, New Stockwell House, Stockwell Street, Leek ST13 6HQ.

Staffordshire Moorlands Footpath Guide. Twelve leaflets, showing walks based on public rights of way, some of which link with the canal. Information from the Planning Officer as above.

Coventry Canal

Coventry's Waterway – A City Amenity, from The Coventry Canal Society, 64 Broad Lane, Coventry, CV5 7AF.

Alvecote Pools Nature Trail, from Warwickshire Nature Conservation Trust Limited, 1 Northgate Street, Warwick CV34 4SP.

Cromford Canal

A Year on the Cromford Canal, a survey by Matlock Field Club from Mrs E. Thorpe, Heatherbank, Smedley Street West, Matlock, Derbyshire.

The Birds of the Cromford Canal, from Cromford Canal Society, Old Wharf, Cromford, Derbyshire.

River Etherow

Etherow Country Park – Keg Trail, from Metropolitan Borough of Stockport, Memorial Park, Marple, Stockport, Cheshire SK6 6BB.

River Foss

The River Foss, from River Foss Amenity Society, 16 The Greenway, Holly Tree Meadows, Haxby, York YO3 8FE.

Glamorgan Canal

Visitor Guide to Glamorgan Canal Local Nature Reserve Whitchurch, from Cardiff City Council, Leisure and Amenities Department, Heath Park, Cardiff.

Gloucester and Sharpness Canal

Severnside: A Guide to Family Walks, edited by Stephen Taylor; published by Croom Helm.

Grand Union Canal (including some of the branches)

Tring Reservoirs Nature Trail, from Nature Conservancy Council, P.O. Box, Godwin House, George Street, Huntingdon PE18 6BU.

South East England – A Guide to Family Walks. Edited by Ian Campbell; published by Croom Helm includes two trails (history and general interest) on the canal at Denham and Aldbury.

Waterside Walks in Northamptonshire, published by John Anderson, 29 The Fairway, Blaby, Leicester.

Waterside Walks in Leicestershire, published by John Anderson, as above.

Out and About in Northamptonshire, from Leisure Services Department, Northamptonshire County Council, Northampton House, Northampton.

Old Union Canals Society Guide to the Union Waterways of Leicestershire and Northamptonshire, from M. Bindley, 32 Coventry Road, Market Harborough, Leics.

Walks in Milton Keynes (a pack containing leaflets, several of which feature the canal); *Manor Farm Trail,* from Milton Keynes Development Corporation, Wavenden Tower, Wavenden, Milton Keynes, Bucks.

Grand Western Canal

Devon Wetlands, published by Devon County Council, Topsham Road, Exeter.

River Granta

Paradise Nature Trail; and *Coe Fen,* both available from Cambridge City Council.

Grantham Canal

Flowers Along the Grantham Canal, from K. Brockway, 473 Woodborough Road, Nottingham NG3 5FR.

River Heddon

Devon Wetlands, published by Devon County Council, Topsham Road, Exeter.

Huddersfield Canals

Huddersfield Narrow Canal – Tame Valley Trail, from Tameside Education Committee, Education Office, Town Lane, Dukinfield SK16 4BY.

The Huddersfield Canals – Towpath Guide, from the Huddersfield Canal Society, Mrs Buckley, 37 Edward Street, Oldham.

Kennet and Avon Canal

Widcombe and Bath Nature Trail, from Department of Leisure and Tourist Services, Bath City Council, Pump Room, Bath BA1 1LZ.

Severnside – A Guide to Family Walks, edited by Stephen Taylor; published by Croom Helm.

Lancaster Canal

Field End Bridge to Stainton: A Canal Bank Nature Trail, from Cumbria Trust for Nature Conservation.

River Lee Navigation

South East England: A Guide to Family Walks, edited by Ian Campbell; published by Croom Helm.

Discovering the Bow Back Rivers, from J. Nicholas, 300 Baring Road, London SE12

Leeds and Liverpool Canal

Jay Nature Trail, from Art Gallery and Museum, Cliffe Castle, Keighley, Yorkshire.

Along T'Cut: A Towpath Trail through Burnley, from Burnley Central Library, Grimshaw Street, Burnley, Lancashire.

Towpath Trail: Between Cherry Tree and Tarleton, from Guardian Press, 32a Market Street, Chorley, Lancashire.

Llangollen Canal (The Welsh Section of the Shropshire Union Canal)

Sandstone Trail Guide, from Cheshire County Council, County Hall, Chester CH1 1SF. A Teachers' Pack is also available.

Prees Branch Nature Reserve, from Shropshire Conservation Trust, Bear Steps, Shrewsbury SY1 1UH

Macclesfield Canal

Bollin Valley Teachers' Pack, from Cheshire County Council, County Hall, Chester CH1 1SF

River Nene

Seeing Northamptonshire's Wildlife, discusses ten sites, including Thrapston Gravel Pits and Titchmarsh Heronry. Available from Northamptonshire Naturalists' Trust Limited, Lings House, Billing Lings, Northampton NN3 4BE.

The River Nene in Northamptonshire, from Leisure Services Department, Northamptonshire County Council, Northampton House, Northampton NN1 2JP

River Great Ouse

Riverside Walk, Stony Stratford. One of the walks in Milton Keynes pack of leaflets, from Milton Keynes Development Corporation, Wavenden Tower, Wavenden, Milton Keynes, Bucks.

Ely Nature Trail, from Cambridgeshire and Isle of Ely Naturalists' Trust Limited, 1 Brookside, Cambridge, CB2 1JF.

Ouse Washes – on Old and New Bedford Rivers, see under Bedford Rivers.

Peak Forest Canal

Brabyns Park Interest Trail, from East Area Office, Administrative Division, Metropolitan Borough of Stockport, Memorial Park, Marple, Stockport, Cheshire SK6 6BB.

Tame Valley Canal Trails, from Chief Tame Valley Warden, Tameside Metropolitan Borough Council, Town Hall, Denton, Manchester.

River Severn

Severnside – A Guide to Family Walks, edited by Stephen Taylor; published by Croom Helm.

Shropshire Union Canal (see also the Llangollen Canal)

Market Drayton Nature Trail, from Shropshire Conservation Trust, Bear Steps, Shrewsbury SY1 1UH.

The Birds of Belvide, (feeder reservoir near Brewood), from A. Richards, 1 Lansdowne Road, Studley, Warwickshire.

Staffordshire and Worcestershire Canal

Valley Park, Wolverhampton, Nature Trail, from Environmental Department, Wolverhampton Metropolitan Borough Council, Civic Offices, Wolverhampton.

River Soar

Walks Around Loughborough, Discovering Charnwood – Walks Around Barrow-on-Soar No 5, from Charnwood Community Council, John Storer House, Wards End, Loughborough.

Stratford-upon-Avon Canal

Earlswood Nature Trail, from Museum and Art Gallery, Congreve Street, Birmingham 3.

Lapworth to Lowsonford Nature Trail, from County of Warwick Museum, Market Place, Warwick.

River Taf

Taf Fechan Nature Reserve – A Guided Walk, from Merthyr Tydfil Borough Council, Town Hall, Merthyr Tydfil, Mid-Glamorgan.

Trent and Mersey Canal

Marbury Country Park, information from Director of Countryside and Recreation, Cheshire County Council, County Hall, Chester CH1 1SF.

The Staffordshire Way, from the County Planning Department, Staffordshire County Council, Martin Street, Stafford ST16 2LE.

Trent Navigation (River Trent)

Attenborough Nature Trail, from the Nottinghamshire Trust for Nature Conservation, Shire Hall, High Pavement, Nottingham NG1 1HR.

Walks in Nottinghamshire, No 42, Starting at Gunthorpe, from the Planning and Transportation Department, Nottinghamshire County Council, Trent Bridge House, Fox Road, West Bridgford, Nottingham NG2 6BJ.

Tavistock Canal

Devon Wetland, published by Devon County Council, Topsham Road, Exeter.

Morwellham Quay Blue Trail, from Morwellham Recreation Centre, Morwellham, near Tavistock, Devon.

Union Canal, Scotland

Wild Plants of the Canal; Birds of the Canal. Both are leaflets in a pack of eight available from British Waterways Board, Old Basin Works, Applecross Street, Glasgow C4.

River Wear

Cathedral Peninsula Nature Trail, from Durham County Conservation Trust, 52 Old Elvet, Durham DH1 3HN.

Witton-Le-Wear Nature Trail, from Durham County Conservation Trust as above.

Walks from Willington Lido Picnic Site; Stanhope Riverside Walk, from Wear Valley District Council, Old Bank Chambers, 45 Market Place, Bishop Auckland, Co Durham.

References

Rivers and Canals

British Waterways Board (1965)	*The Facts about the Waterways*	(British Waterways Board)
Clapham, Sir John (1957)	*A Concise Economic History of Britain from the earliest times to 1750*	(CUP)
Hadfield, C. (1968)	*The Canal Age*	(David & Charles)
Hadfield,C. (1973)	*British Canals*	(David & Charles)
Hoskins, W.G. (1976)	*The Age of Plunder: The England of Henry VIII (1500-47)*	(Longman & Co)
McKnight, H. (1975)	*The Shell Book of Inland Waterways*	(David & Charles)
Murphy, B. (1973)	*A History of the British Economy 1086-1740*	(Longman)
Nicholson, R. (undated)	*Nicholson's Guide to the Waterways, 5: Midlands*	(Robert Nicholson's Publications)

Animals

Askew, R. R. (1971)	*Parasitic Insects*	(Heinemann)
Beirne, B. P. (1952)	*British Pyralid and Plume Moths*	(Warne)
Clegg, J. (1974)	*Freshwater Life*	(Warne)
Corbet, G. B. and Southern, H. N. (1977)	*Handbook of British Mammals*	(Blackwell)
Corbet, P., S., Longfield, C. and Moore, N. W. (1960)	*Dragonflies*	(Collins)
Cramp, S. et al	*Birds of the Western Palearctic, Vols 1 and 2*	(Oxford)
Elliott, J. M. (1977)	*Megaloptera and Neuroptera*	(Freshwater Biological Association)
Elliot, J. M. and Mann, K. H. (1979)	*A Key to the British Freshwater Leeches*	(Freshwater Biological Association)
Hammond, C. O. (1977)	*Dragonflies of Great Britain and Ireland*	(Curwen)
Heath, J. (1979)	*Moths and Butterflies of Great Britain and Ireland : 9*	(Curwen)
Hickin, N. (1952)	*Caddis*	(Methuen)
Hynes, H. B. N. (1971)	*The Biology of Polluted Waters*	(Liverpool University Press)
Hynes, H. B. N. (1970)	*The Ecology of Running Water*	(Liverpool University Press)
Hynes, H. B. N. (1967)	*Stoneflies*	(Freshwater Biological Association)
James A. and Evison, L. (1979)	*Biological Indicators of Water Quality*	(Wiley)

Jones, J. R. G. (1964)	*Fish and River Pollution*	
Macan, T. T. (1973)	*British Trichoptera*	(Freshwater Biological Association)
Macan, T. T. (1979)	*Nymphs of British Ephemeroptera*	(Freshwater Biological Association)
Maitland, P. S. (1977)	*Coded Checklist of Animals Occurring in Freshwater in the British Isles*	(Natural Environment Research Council)
Marchant, J. H. and Hyde, P. A. (1980)	'Aspects of the Distribution of Riparian Birds on Waterways in Britain and Ireland'	*Bird Study,* **27**, pp. 183-202
Marchant, J. H. and Hyde, P. A. (1980)	'Population Changes for Waterways Birds, 1974-78'	*Bird Study,* **26**, pp. 227-238
Marchant, J. H. and Hyde, P. A. (1980)	'Population Changes for Waterways Birds 1978-79'	*Bird Study,* **27**, pp. 179-182
Matthews, L. H. (1952)	*British Mammals*	(Collins)
Mellanby, K. (1971)	*The Mole*	(Collins)
O'Connor, F. B. (1979)	*Second Report of the Joint Otter Group*	(Nature Conservancy Council/ Society for the Promotion of Nature Conservation)
Oldroyd, H. (1969)	*Diptera-Brachycera: Tabanoidea and Asiloidea*	(Royal Entomological Society)
Richards, O. W. (1956)	*Hymenoptera: Introduction and Key to Families*	(Royal Entomological Society)
Sharrock, J. T. R. (1976)	*Atlas of Breeding Birds in Britain and Ireland*	(British Trust for Ornithology)
Smith, A. E. (1975)	'The Impacts of Lowland River Management'	*Bird Study,* **22**, pp. 249-54
Southwood, T. R. E. and Leston, D. (1959)	*Land and Water Bugs*	(Warne)
Whitton, B. A. (1975)	*Studies in Geology 2 – River Geology*	(Blackwell)
Willoughby, L. G. (1976)	*Freshwater Biology*	(Hutchinson)
Witherby, H. F. et al (1938-41)	*Handbook of British Birds (Vols 1-5)*	(Witherby)
Youngman, R. E. (1977)	'Great Crested Grebes Breeding on Rivers'	*British Birds,* **70**, pp. 544-5

Habitats and Plants

Belcher, H. and Swale, E. (1976)	*A Beginners Guide to Freshwater Algae*	(HMSO)
Belcher, H. and Swale, E. (1979)	*An Illustrated Guide to River Phytoplankton*	(HMSO)
Bibby, C. J. and Lunn, J. (1981)	'Conservation of Reedbeds and their Avifauna in England and Wales'	*Biological Conservation* (in press)
Brightman, F. H. and Nicholson, B. E. (1966)	*The Oxford Book of Flowerless Plants*	(OUP)
Clapham, A. R., Tutin, T. G. and Warburg, E. F. (1981)	*Excursion Flora of the British Isles* (Third edition)	(CUP)
Engelhardt, W. (1973)	*The Young Specialist Looks at Pond Life*	(Burke)
Fitter, R. S. R. (1971)	*Finding Wild Flowers*	(Collins)
Haslam, S. M. (1978)	*River Plants*	(CUP)
Haslam, S. M., Sinker, C. and Wolseley, P. (1975)	'British Water Plants'	*Field Studies,* **4**, pp. 243-351
Jermy, A. C. and Tutin, T. G. (1968)	*British Sedges*	(Botanical Society of the British Isles)
Keble Martin, W. (1969)	*The Concise British Flora in Colour*	(Ebury Press/Michael Joseph)
McClintock, D. and Fitter, R. S. R. (1956)	*Collins Pocket Guide to Wild Flowers*	(Collins)
Maitland, P. S. (1978)	*Biology of Fresh Waters*	(Blackie)
Perring, F. (ed) (1970)	*The Flora of a Changing Britain*	(Pendragon)

Perring, F. and Walters S. M. (1976) *Atlas of the British Flora* (Botanical Society of the British Isles)

Pollard, E., Hooper, M. D. and Moore, N. W. (1974) *Hedges* (Collins)

Popham, E. J. (1955) *Some Aspects of Life in Fresh Water* (Heinemann)

Rose, F. (1981) *The Wildflower Key* (Warne)

Sculthorpe, C. D. (1967) *The Biology of Aquatic Vascular Plants* (Arnold)

Townsend, C. R. (1980) *The Ecology of Streams and Rivers Studies in Biology: 122* (Arnold)

Twigg, H. M. (1959) 'Freshwater Studies in the Shropshire Union Canal' *Field Studies,* **1**, No 1

Conservation Problems

British Waterways Board (1975) *Recreational Use of Inland Waterways: Data 1967-74* (British Waterways Board)

Haslam, S. M. (1973) 'The Management of British Wetlands 1 — Economic and Amenity Use' *Journal of Environmental Management,* **1**, pp. 303-320

Inland Waterways Amenity Advisory Council (1975) *Angling on the British Waterways Board System* (IWAAC)

Inland Waterways Amenity Advisory Council (1978) *Recreation on Inland Waterways* (IWAAC)

Liddle, M. J. and Scorgie, H. R. A. (1980) 'The Effects of Recreation on Freshwater Plants and Animals: A Review' *Biological Conservation,* **17**, pp. 183-206

Mellanby, K. (1967) *Pesticides and Pollution* (Collins)

Mellanby, K. (1972) *The Biology of Pollution* *Studies in Biology, No 38, (Arnold)*

Nature Conservancy Council (1981) *Lead Poisoning in Swans* (Nature Conservancy Council)

O'Riordan, T. and Paget, G. (1978) *Sharing Rivers and Canals, Sports Council Study 16* (Sports Council)

Perring, F. H. and Farrell, L. (1977) *British Red Data Book 1: Vascular Plants* (Society for the Promotion of Nature Conservation)

RSPB/SPNC/Wessex Water Authority (1980) *Rivers and Wildlife* (SPNC)

Salter, J. H. (1894) *A Guide to the Thames and Principal Canals* (Alden & Co, London)

Tanner, M. F. (1973) *Water Resources and Recreation: Study 3* (Sports Council)

Water Space Amenity Commission (1980) *Conservation and Land Drainage Guidelines* (WSAC)

INDEX